Studies in Systems, Decision and Control

Volume 184

Series editor

Janusz Kacprzyk, Polish Academy of Sciences, Warsaw, Poland
e-mail: kacprzyk@ibspan.waw.pl

The series "Studies in Systems, Decision and Control" (SSDC) covers both new developments and advances, as well as the state of the art, in the various areas of broadly perceived systems, decision making and control–quickly, up to date and with a high quality. The intent is to cover the theory, applications, and perspectives on the state of the art and future developments relevant to systems, decision making, control, complex processes and related areas, as embedded in the fields of engineering, computer science, physics, economics, social and life sciences, as well as the paradigms and methodologies behind them. The series contains monographs, textbooks, lecture notes and edited volumes in systems, decision making and control spanning the areas of Cyber-Physical Systems, Autonomous Systems, Sensor Networks, Control Systems, Energy Systems, Automotive Systems, Biological Systems, Vehicular Networking and Connected Vehicles, Aerospace Systems, Automation, Manufacturing, Smart Grids, Nonlinear Systems, Power Systems, Robotics, Social Systems, Economic Systems and other. Of particular value to both the contributors and the readership are the short publication timeframe and the world-wide distribution and exposure which enable both a wide and rapid dissemination of research output.

More information about this series at http://www.springer.com/series/13304

Peter Simon Sapaty

Holistic Analysis and Management of Distributed Social Systems

 Springer

Peter Simon Sapaty
Institute of Mathematical Machines
  and Systems
National Academy of Sciences
Kiev, Ukraine

ISSN 2198-4182   ISSN 2198-4190   (electronic)
Studies in Systems, Decision and Control
ISBN 978-3-030-13197-5   ISBN 978-3-030-01830-6   (eBook)
https://doi.org/10.1007/978-3-030-01830-6

This Springer imprint is published by the registered company Springer Nature Switzerland AG
The registered company address is: Gewerbestrasse 11, 6330 Cham, Switzerland

*In memory of my father Simon, great
biologist and scientific believer*

Preface

What written in this book, has started from the end of sixties of the previous century with our participation in the creation of citywide distributed heterogeneous computer networks (well before the internet), which were integrating quite different machines installed in computer centers located far away from each other and communicating via telephone lines. There was a need to solve complex numerical-analytical problems on such networks, which could not be done efficiently on any single machine, as some were working directly with mathematical formulae in hardware-emulated language ANALYTIC, while others engaged in numerical computing under compiled FORTRAN.

At that time creating some higher-level language with its implementation covering the whole application area on each computer could be very expensive, even impossible, which prompted a crazy at first sight idea to leave everything as it is but rather express numerical-analytical algorithms as integrated patterns in a mixture of the two languages. These higher-level patterns were capable of moving through and matching the network, being materialized into concrete language-oriented stages when reaching proper computers. For example, formulae obtained in ANALYTIC in one computer were automatically inserted into this pattern with its further materialization into a FORTRAN stage for numerical computations on another computer.

Despite oriented on specific applications, this idea of representing solutions in distributed computer networks as *moving and matching patterns* rather than traditional parts-agents exchanging messages, appeared to be extremely productive for many subsequent networking applications in both civil and defence areas, which also resulted in a European patent and previous Wiley and Springer books. A special language, called WAVE, first appeared for formal expression of such spatial patterns wavelike covering distributed networks, in which a completely new style of parallel spatial programming began developing. Its different successful implementations helped us to cover more and more application areas under such pattern matching philosophy which, in its turn, allowed the very language to be gradually enriched and improved, with its latest version used in this book called as *Spatial Grasp Language*.

Programming in such spatial pattern-matching style in distributed spaces appeared to be psychologically close to our mental perception and solutions of different problems in distributed spaces, where we *do not think of any computers or communications between them* but rather feeling *direct presence* in these spaces with full *freedom of movement* through them. This style also happened to be very close to the gestalt psychology and theory born in Germany a century ago, which shows that human mind is capable of directly perceiving the whole as patterns, with parts treated in the context of this whole rather than vice versa.

The current book, as a sequel to the previous ones, especially recent Springer book, is considering a new application of the developed paradigm in a broad and highly contested area as large social systems and their representation as social networks. These are actually forming the living tissue of humankind, with expected influence of the current book in many related areas, like national and international economic and industrial systems, culture, security, welfare, and others—of course, being the author's dream at this initial stage.

Kiev, Ukraine   Peter Simon Sapaty

Acknowledgements

To the following persons and organizations who supported this book.

John Page, University of New South Wales, Australia, for numerous discussions related to unmanned systems, collective behavior, and social robotics, also cooperation within the board of International Journal of Intelligent Unmanned Systems (ISSN: 2049-6427).

Aidan Morrison and Martha Musandu of Emerald Publishing, for useful advices and moral support of this book, also Emerald Literati Team for the recent award of author's paper mentioned in the book.

Springer International Publishing, for years of publication of the author's book chapters and recent full book, and personally Thomas Ditzinger, Lavanya Diaz, and Prasanna Kumar Narayanasamy for high professionalism and valuable recommendations.

International Relations and Diplomacy (ISSN: 2328-2134) and personally Melian Lee, for the support of author's publications mentioned in this book and valuable recommendations on paper preparations, also fruitful cooperation within the editorial team.

Masanori Sugisaka, ALife Robotics, Japan, with whom the author has long and fruitful cooperation in the area of Artificial Life and Robotics, organization of international ALIFE conferences, also having common publications with the author mentioned in the book.

Takao Ito, Hiroshima University, Japan, with whom we started current book-related fruitful cooperation in the area of large social systems and networks, and especially linked with industrial Keiretsu networks in Japan mentioned in this book, with common publications planned.

Stephen Lambacher from School of Social Informatics, Aoyama Gakuin University in Tokyo Japan, a great linguist, with whom the author discussed many original ideas preceding this book.

Robert Finkelstein, Robotic Technology Inc., USA, with whom the current and previous books-related ideas were frequently discussed, and particularly in the areas of human terrain, psychology, philosophy, robotics, and driverless transport.

Bob Nugent, retired US Navy Commander and now with Ph.D. at Virginia Tech, USA with whom the author had meetings and useful discussions on advanced command and control, which appeared useful for social systems discussed in this book.

As the current book has links and growing plans oriented on large distributed systems, economic ones including and especially, very useful discussions on that matter were with my grandson Eugene majoring in international economy as final year university student.

And gratitude to my wife Lilia courageously tolerating the fourth book writings and closing eyes on missed holidays, while supporting the necessary links and contacts with the real world, with myself 24/7 existing in the virtual one during final book preparations.

Kiev, Ukraine   Peter Simon Sapaty

Contents

About the Author

Dr. Peter Simon Sapaty, Chief Research Scientist, Ukrainian Academy of Sciences, is with networked systems for five decades. Except Ukraine, worked in Czechoslovakia, Germany, UK, Canada, and Japan as group leader, Alexander von Humboldt researcher, and special invited, invited and visiting professor, also created and chaired SIG on Mobile Cooperative Technologies within Distributed Interactive Simulation project in the US. Invented distributed control technology resulted in European patent and Wiley and Springer books, published more than 200 papers on distributed system organizations. Regularly served as workshop organizer, sessions chair, keynote and invited speaker at scientific conferences in different countries. His bio is in Marquis Who's Who in the World and Cambridge Outstanding Intellectuals of the 21st Century. Peter is also engaged in different international scientific journals, as editor-in-chief including.

Chapter 1
Introduction

1.1 What Is This Book About?

This book is the fourth in line devoted to seeing, understanding, processing, and management of large distributed networked systems, stemming from the author's work on design and implementation of heterogeneous citywide computer networks at the end of sixties, well before the internet.

The developed high-level approach of dealing with distributed networked structures is describing solutions in them in the form of *active spatial patterns* that can be applied from any network locations and then *self-spread*, *self-replicate*, *self-cover* and *self-match* the network topologies while dynamically creating runtime distributed operational infrastructures in the networks for solving any problems *over*, *on*, *in*, *under*, *against*, and *about* them.

This holistic paradigm can effectively grasp complex networked solutions on topmost level and in transparent and compact manner while ideologically, psychologically and technologically *starting from the whole* rather than from its parts. This is fundamentally different from traditional models and organizations attempting to get the desired and often miraculous whole starting from predetermined parts-agents and their interactions. This invented, patented, developed, and prototyped approach has been successfully used for solving networked problems in a variety of areas, both civil and defence, and in different countries.

The current book represents its further development with orientation on solving problems in distributed social systems and their formal representation as social networks. Such systems can be very large and dynamic, with changeable relations between parts which cannot, due to existing processing, storage and visualization capabilities, even in principle, be copied, analyzed, and understood in centralized locations. The technology developed is allowing them to be effectively dealt with in fully distributed, parallel, and cooperative manner, directly in the locations where dynamic network items are residing, and with the use of any available computation and communication resources throughout such systems.

© Springer Nature Switzerland AG 2019   1
P. S. Sapaty, *Holistic Analysis and Management of Distributed Social Systems*, Studies
in Systems, Decision and Control 184, https://doi.org/10.1007/978-3-030-01830-6_1

Under the paradigm developed, the book also considers examples of effective dealing with such modern trends as robotization and massive use of driverless cars, which are becoming inseparable parts and features of advanced social systems, enriching their semantics, dynamics, prosperity, and security. And finally, the holistic nature of the approach discussed will be demonstrated as ideologically and conceptually linked with the Gestalt psychology and theory whose main ideas and laws are usually demonstrated on localized visual images perceived by biological brain only. In our case, such laws can also be effectively extended and applied to the distributed networked systems, social ones including, under the developed high-level spatial intelligence, with further research planned in this challenging area, beyond the scope of the current book.

1.2   The Century of Large Scale Systems

The whole world is steadily moving towards quite different system organizations not seen and described before, which are sometimes called *Ultra-Large Scale Systems* [1]. Such systems include people, organizations, and technologies at all levels with significant and often competing interdependencies. They are continuously adapting in different points where inherent conflicts and failures being norms rather than exceptions. Their main features, discussed in [1], may be summarized as follows:

- Decentralization
- Inherently conflicting, unknowable, and diverse requirements
- Continuous evolution and deployment
- Heterogeneous, inconsistent, and changing elements
- Erosion of the people/system boundaries
- Normal failures
- New paradigms for acquisition and policy emerging
- Rapidly increasing systems complexity.

Such organizations relate to *ultra-large, network-centric, real-time, cyber-physical-social systems* having thousands of platforms, sensors, decision nodes, which are connected through heterogeneous wired and wireless networks. Another class, *socio-technical ecosystems*, includes people, organizations and technologies at all levels with significant and often competing interdependencies. Among numerous societal problems in ultra-large systems, such as climate change and the environment, powering of societies, disease, epidemics, health care, growing megacities, safety and security, transportation, and many others can be named.

Large-scale distributed systems have become an integral part of everyday's life, investigation and research, with the development of large web applications, social networks, peer-to-peer systems, wireless sensor networks and so on. For example, the research group at Vrije University in Amsterdam [2] addresses numerous questions related to the way we may design, build, operate and maintain such large-scale systems.

Radically new models and technologies for such large distributed systems are needed for maintaining their stability, prosperity, safety, or even survivability in different (including emergency and asymmetric) situations reflecting and responding to the rapidly growing world dynamics. Further development of the distributed control ideology and technology pursued for decades with new applications to large distributed social systems is the main purpose of this book.

1.3 Social Systems and Networks

In sociology, a *social system* is the patterned network of relationships constituting a coherent whole that exist between its different elements [3]. In another wording, it may be considered as a bounded set of interrelated activities that together constitute a single entity. A social system is composed of persons or groups of persons who interact and mutually influence each other's behavior. Social systems exist at all levels: persons, families, organizations, communities, societies, cultures.

Examples of social systems usually include nuclear family units, communities, cities, nations, college campuses, corporations, and industries [4]. The organization and definition of groups within a social system depend on various shared characteristics such as location, socioeconomic status, race, religion, societal function, or other distinguishable features.

Social systems have been studied for as long as sociology has existed, with numerous publications emerging in this area, with examples in [5–8]. Talcott Parsons is often considered the first to formulate a systematic theory of social systems [5], having defined a social system as a network of interactions between actors. According to Parsons, social systems rely on a system of language, and culture must exist in a society in order to qualify it as a social system. His work laid the foundations for the rest of the study of social systems theory and ignited the debate over what framework social systems should be built around, such as actions, communication, or other relationships.

Social systems are usually represented by *social networks* [9–15], with the latter formulated as a network of individuals such as friends, acquaintances, or co-workers connected by interpersonal relationships. Social networks are also often referred to as online service or site through which people create and maintain interpersonal relationships, as in [9]. Usually, social networks are represented as graphs with nodes reflecting individuals or organizations linked with each other by one or more arcs, the latter expressing specific types of interdependency such as friendship, kinship, common interest, financial exchange, dislike, sexual relationships, or relationships of beliefs, knowledge or prestige.

Social networks are extensively analyzed and investigated [10–13], with the resulting graph-based structures being often very complex and having many kinds of ties between the nodes. Research in a number of academic fields has shown that social networks operate on many levels, from families up to the level of nations, and play a critical role in determining the way problems are solved, organizations are run, and the degree to which individuals succeed in achieving their goals.

People have used the idea of "social network" loosely for over a century to connote complex sets of relationships between members of social systems at all scales, from interpersonal to international. Rather than treating individuals (persons, organizations, states) as discrete units of analysis, it focuses on how the structure of ties affects individuals and their relationships. Social network analysis produces an alternate view, where the *attributes of individuals are less important than their relationships and ties* with other actors within the network.

Analysing social networks can help in many areas of human activity, like, for example, in law enforcement [14]. Social networks also play a key role in hiring, in business success, and in job performance [15]; they usually provide ways for companies to gather information, deter competition, and collude in setting prices or policies.

Nowadays, the huge amounts of data available in relation to social networks pose serious problems for analysis with regular hardware and/or software [13]. Emerging technologies, like modern models for parallel computing, multicore computers or clusters of computers, can be very useful for analyzing massive network data. The current book is just in line with these demands and trends by offering a fully distributed ideology and technology for dealing with networks without limitations on their volume, complexity, and distribution in both virtual and physical spaces. This technology allows for potential use of any computational and communication resources distributed or scattered in large physical spaces occupied by the networks, which can be engaged at runtime collectively, cooperatively, and in parallel.

1.4   The Technology Development History

We are providing here a brief history of the developed network processing approach being the basis of the current and previous books.

The author's networking experience started half century ago by first dealing with large distributed power networks and calculating their regimes and distribution of energy (while still being a student). It then continued in a quite different but also networking area—by creating from the end of sixties the first citywide heterogeneous computer networks. The operational scenarios in such networks for solving (complex numerical-analytical) problems were written in an integrated mixture of very different, specialized, programming languages that could self-navigate and cover networks while organizing collective task solutions in them [16–25]. That unusual organization allowed us to avoid creation of a single and expensive at that time universal language and its implementation acceptable by all (very different, even incompatible) computers in the same network.

The spatial navigation of networks by integral scenario code, which allowed us to drastically simplify distributed algorithms and their implementation in comparison with approaches based on communicating agents, was then extended to working with distributed semantic networks and developing a special higher-level language for this, along with its distributed interpretation systems [26–37].

Further activities in this area aimed at using the developed networking paradigm for distributed command control in a variety of systems [38–42], spatial intelligence, dominance, and integrity [43–55], distributed simulation and management of large dynamic systems in defense [56–68], human-robotic systems and collective robotics [69–78], societal systems and solving problems in them [79–81]. Other activities and investigations had been linked with integrated air and missile defense [82, 83], gestalt-psychology and theory [84], crisis management [85], night vision management [86], directed energy systems [87–89], and among the latest, with driverless transport and road management with it [90]. The approach developed, known in the past as WAVE, had trial implementations is different countries and was in public domain on the internet.

These mentioned works also resulted in a European patent [91] and Wiley and Springer books [92–94], with the current one actually being a sequel to them. Many more details on the history and background of the developed approach can be found in the existing publications, especially in the preceding Springer book [94]. So we will not be repeating them here, only providing the technology's latest state of the art and brief summary in the following chapters, fully sufficient for understanding the rest of the book.

1.5   How the Book Is Organized

The rest of the book is organized as follows.

Chapter 2 contains summary of the developed Spatial Grasp Model and Technology (SGT) based on solving problems in arbitrary large distributed systems by high-level active patterns which can self-propagate, self-replicate, self-modify, and self-match such systems in a controlled flooding or super-virus mode, without central resources. Numerous communicating interpreters from its basic Spatial Grasp Language (SGL), collectively executing SGL scenarios, can be installed in key system points (on agreements or in a stealth manner for special applications). SGT allows us to start from any system points and dynamically cover any regions of interest using unlimited scenario-pattern mobility. Any existing social media tools and platforms can be extended for having this capability by supplying them with implanted SGL interpreters. The latter can also be installed in any human-wearable devices and robotic units, allowing them all operate in integral teams.

Chapter 3 describes full details of the latest version of SGL which can be particularly suitable for dealing with very large distributed social systems and social networks representing them. Extremely compact SGL operational scenarios (usually hundreds of times shorter than in C or Java as pursuing quite different philosophy of dealing with distributed spaces) are represented as active recursive patterns. The latter are directly, spatially matching distributed systems rather than copying and bringing them for analysis, visualization, and processing to separated computational resources, as usual. SGL allows us to create and process large and complex graphs and networks that can be arbitrarily distributed in both virtual and physical spaces, with the use of all computational and communication facilities available throughout this space. SGL has unified recursive syntax and semantics with clear interpretation scheme allowing it to be easily implemented for concrete applications as its any subset or, on the opposite, extension.

Chapter 4 is devoted to expression and explanation of distributed network processing basics in SGL which are all oriented on parallel and fully distributed dealing with networks of arbitrary size, structure, and space coverage. For demonstrations, the kite-like network topology widely available on internet is used, just for making tasks formulated and solutions offered more convenient and understandable to possible broader audience. The SGL solutions include initial direct setting on network nodes, combination of stepwise and parallel movements through the network structure, calculation of numbers of distributed network elements, forming breadth-first and depth-first spanning trees in the networks, finding different paths through network nodes and links, shortest ones including, as well as initial creation and modification of arbitrary networks in distributed spaces.

Chapter 5 offers, in SGL too, detailed solutions for discovering and analyzing basic and widely recognized features and parameters of social networks, using for demonstrations another well known network topology available on the internet (for same reasons as in the previous chapter). These include, first of all, the traditional centrality issues like degree centrality, eigenvector centrality, closeness centrality, and betweenness centrality of its nodes, with finding best candidates among them for these roles. Also considered are network clustering, finding both strongest subnetworks (like cliques) and weakest (or articulation points), recognizing particular structures with the use of arbitrary complex search and matching patterns, which may contain variants and alternatives. Others solutions are dealing with assessment of physical parameters (if available) of distributed social networks, including finding topological centers of different communities. This, for example, may be useful for prediction or/and prevention of possible conflicts between them, or on the opposite, for improving social integration of these communities, say, into one national system.

Chapter 6 is trying to deal with some real and complex networks using the developed and tested mechanisms and solutions of the previous two chapters. These potentially distributed network structures reflect industrial organizations common in Japan like Keiretsu networks (having interwoven combination of vertical and horizontal features). Others relate to large international networks expressing worldwide business ecosystems with different types of nodes and investment and/or strategic partnership links between them, also academic networks showing interactions between persons with different scientific degrees. Exemplary operations on such types of networks are offered, confirming transparency, compactness, and simplicity of SGL scenarios, where each network node can be located in a different virtual, physical, or computing place. Typical operations on very large networks with numerous nodes and links, looking altogether as "hairball" on a single picture if somehow collected there, and about which nothing is known in advance, are presented in SGL too.

Chapter 7 is offering a higher or "over-operability" layer for human-robot integration based on SGT, rather than following traditional "interoperability" models having serious weaknesses when managing large dynamic and distributed systems in trying to pursue global goals. This layer allows us to express top semantics of main operations and decisions in distributed spaces regardless of being performed by humans or robots which can substitute each other at any time. A related tasking and organization of human-robotic teams at different levels for a combined territory search is shown. As autonomous vehicles are becoming inseparable part of social systems, the chapter presents some typical collective scenarios on roads with massive use of driverless cars, under the over-operability philosophy too. The shown SGL scenarios include narrowing of gaps before vehicles, lane manoeuvring, collective avoidance of broken cars, effective platoon management, as well as autonomous distributed routing in road networks by manned or unmanned cars, the latter also capable of chasing each other under the help of radar networks aggregated with road infrastructures.

Chapter 8 is linked with the Gestalt psychology and theory emphasizing the unique capability of human mind and brain to directly grasp complex images as a whole while interpreting their parts, which may be incomplete, in the context of this whole rather than vice versa. Using SGT, we can extend and extrapolate these originally single-brain gestalt features from traditional visual domain to seeing and understanding structures and situations distributed throughout large spaces and, moreover, do this remotely and in parallel. The chapter shows how main gestalt principles, often referred to as laws, can be expressed and tested on images placed on distributed canvas represented, for convenience, by regular networks—with potential importance of the gained experience for holistic analysis and management of large distributed systems, social ones including. The related demonstration in SGL covers such gestalt laws as proximity, similarity, continuity, closure, common fate, symmetry, past experience, and good gestalt, as well as figure/ground controversy and connectedness or the law of unity.

Chapter 9 concludes the book, providing summary of its main results which include description of the latest version of the networking technology and numerous solutions on social networks, also related to social robotics, as well as extension of gestalt theory principles to distributed systems. The chapter summarizes the main technology features on a very simple but representative example of finding paths in a network, discusses minimal SGL subset for quick implementation on different platforms. The expected future investigation and research areas include real economic and industrial systems and networks with their distributed interactive simulation, and possible reassessment of classical systems theory in light of the researched in the current book holistic and gestalt principles. The further plans may also include continuing development of the previously published works on wholeness, goal orientation, robustness, and integrity of large systems of different natures. Further publications, books including, in these and other areas are planned too, another technology patenting and subsequent marketing as well.

References

1. L. Northrop, *Ultra-Large-Scale Systems. The Software Challenge of the Future*. Carnegie Mellon, Pittsburgh, PA 15213-3890, June 2006. https://resources.sei.cmu.edu/asset_files/Book/20 06_014_001_30542.pdf
2. Large-scale distributed systems. Vrije Universiteit Amsterdam, 2018. https://www.cs.vu.nl/e n/research/computer-systems/LSDS/index.aspx
3. Definition of social system. https://www.merriam-webster.com/dictionary/social%20system
4. Social system. https://en.wikipedia.org/wiki/Social_system
5. T. Parsons, *The Social System* (Routledge, England, 1951). http://home.ku.edu.tr/~mbaker/CS HS503/TalcottParsonsSocialSystem.pdf
6. J.W. Forrester, Counterintuitive behavior of social systems. Technol. Forecast. Soc. Change **3**, 1–22 (1971–1972)
7. D. Michailakis, W. Schirmer, Social work and social problems: a contribution from systems theory and constructionism. Int. J. Soc. Welfare **23**(4), 431–442 (2014)
8. Social problems & news topics in the twenty-first century. Online course tutorial SOCY 105: introduction to contemporary social problems. University libraries guide. http://lib.guides.um d.edu/c.php?g=326995&p=2194601
9. Socal network. https://www.merriam-webster.com/dictionary/social%20network
10. Social network analysis. Theory and applications. https://www.politaktiv.org/documents/1015 7/29141/SocNet_TheoryApp.pdf
11. A. Abdalla, S.Y. Yayilgan, in *A Review of Using Online Social Networks for Investigative Activities*, ed. by G. Meiselwitz, SCSM 2014, LNCS 8531 (Springer International Publishing, Switzerland, 2014), pp. 3–12. https://pdfs.semanticscholar.org/bfe6/f640dbe484be95f930bc6 d4fe4b7fb91e4ee.pdf
12. D. Rosen, M.A. Stefanone, D. Lackaff, Online and offline social networks: investigating culturally-specific behavior and satisfaction, in *Proceedings of the 43rd Hawaii International Conference on System Sciences*, 2010. https://www.researchgate.net/publication/224123226_ Online_and_Offline_Social_Networks_Investigating_Culturally-Specific_Behavior_and_Sat isfaction
13. R. Sarmento, J. Gama, A. Bifet, T. Cunha, in *Large Scale Social Network Analysis, Data Analytics 2013, Conference: JOCLAD 2013*, Guimarães. https://www.researchgate.net/public ation/259942858_Large_Scale_Social_Network_Analysis
14. D.S. Polans (ed.), in *Social Network Analysis for Law Enforcement, Standards, Methods, & Technology Committee White Paper 2018*, 02 Feb 2018, International Association of Crime Analysts, IACA. https://www.iaca.net/Publications/Whitepapers/iacawp_2018_02_social_net work_analysis.pdf
15. W. Mayrhofer, Social systems theory as theoretical framework for human resource management—Benediction or curse? Manage. Rev. **15**(2), 178–191 (2004). http://www.hampp-ejour nals.de/hampp-verlag-services/get?file=/frei/mrev_2_2004_178
16. A.T. Bondarenko, S.B. Mikhalevich, A.I. Nikitin, P.S. Sapaty, Software of BESM-6 computer for communication with peripheral computers via telephone channels, in *Computer Software*, vol. 5 (Institute of Cybernetics Press, Kiev, 1970) (in Russian)
17. P.S. Sapaty, A method of organization of an intercomputer dialogue in the radial computer systems, in *The Design of Software and Hardware for Automatic Control Systems* (Institute of Cybernetics Press, Kiev, 1973) (in Russian)
18. J.G. Grigorjev, V.P. Karpus, L.I. Pristupa, P.S. Sapaty, Management of a dialogue in the MIR-2–BESM-6 system, in *Proceedings of Republic Conference on Hardware and Software for Management of Dialogue in Computer Systems*, Kiev, 1973 (in Russian)
19. A.T. Bondarenko, S.B. Mikhalevich, P.S. Sapaty, Intercomputer dialogue in high-level languages, in *Proceedings of Republic Conference on Hardware and Software for Management of Dialogue in Computer Systems*, Kiev, 1973 (in Russian)

20. P.S. Sapaty, *On Possibilities of the Organization of a Direct Intercomputer Dialogue in ANA-LYTIC and FORTRAN Languages*, Publ. No. 74-29 (Institute of Cybernetics Press, Kiev, 1974) (in Russian)
21. P.S. Sapaty, Solving branching and cycling tasks on multiprocessor systems. Proc. USSR Acad. Sci. Tech. Cybern. (1) (1974) (in Russian)
22. P.S. Sapaty, Organization of computational processes in distributed heterogeneous computer networks. Ph.D. Dissertation, Institute of Cybernetics, Kiev, 1976 (in Russian)
23. P.S. Sapaty, On efficient structural implementation of operations on semantic networks. Proc. USSR Acad. Sci. Tech. Cybern. (5) (1983) (in Russian)
24. P.S. Sapaty, Active information field as a model for structural solving of tasks on graphs and networks. Proc. USSR Acad. Sci. Tech. Cybern. (5) (1984) (in Russian)
25. P.S. Sapaty, Solving tasks on semantic networks and graphs by active distributed structures, in *Proceedings of 3rd International Conference Artificial Intelligence and Information-Control Systems of Robots*, Smolenice (Elsevier Science Publishers B.V., North-Holland, 1984)
26. P.S. Sapaty, A wave approach to the languages for semantic networks processing, in *Proceedings of International Workshop on Knowledge Representation. Section 1: Artificial Intelligence*, Kiev, 1984 (in Russian)
27. P.S. Sapaty, A wave language for parallel processing of semantic networks. Comput. Artif. Intell. **5**(4) (1986)
28. P.S. Sapaty, The wave approach to distributed processing of graphs and networks, in *Proceedings of International Working Conference on Knowledge and Vision Processing Systems*, Smolenice, Nov 1986
29. P.S. Sapaty, WAVE-1: a new ideology of parallel processing on graphs and networks, in *Proceedings of International Conference on Frontiers in Computing*, Amsterdam, 1987
30. P.S. Sapaty, The WAVE-1: a new ideology and language of distributed processing on graphs and networks. Comput. Artif. Intell. (5) (1987)
31. P.S Sapaty, Parallel processing for knowledge representation, in *Infotech State of the Art Report on Parallel Processing*, ed. by C. Jesshope (Pergamon Press, England, 1987)
32. Sapaty, P.S., WAVE-1: a new ideology of parallel processing on graphs and networks. Future Generations Comput. Syst. **4** (1988)
33. P.S. Sapaty, The WAVE model for advanced knowledge processing. Report No. OUEL 1803/89, University of Oxford, England, 1989
34. P.S. Sapaty, The WAVE machine project, in *Proceedings of IFIP Workshop on Silicon Architectures for Neural Nets*, St. Paul de Vence, France, 28–30 Nov 1990
35. P.S. Sapaty, Logic flow in active data, book chapter in *VLSI for Artificial Intelligence and Neural Networks* (1991), pp. 79–91
36. P.S. Sapaty, The WAVE paradigm, in *Proceedings of JICSLP'92 Post-Conference Joint Workshop on Distributed and Parallel Implementations of Logic Programming Systems*, Washington, D.C., 13–14 Nov 1992
37. P.S. Sapaty, A brief introduction to the WAVE language. Report No. 3/93, Faculty of Informatics, University of Karlsruhe, 1993
38. P.S. Sapaty, Distributed technology for global control, book chapter in *Informatics in Control, Automation and Robotics*, vol. 37 of the series Lecture Notes in Electrical Engineering (2009), pp. 3–24
39. P.S. Sapaty, Meeting the world challenges with advanced system organizations, book chapter in *Control, Automation and Robotics*, vol. 85 of the series Lecture Notes in Electrical Engineering (2011), pp. 29–46
40. P.S. Sapaty, Grasping spatial integrity in distributed unmanned systems, book chapter in *Informatics in Control Automation and Robotics*, vol. 85 of the series Lecture Notes in Electrical Engineering (2011), pp. 79–97
41. P. Sapaty, Remote control of open groups of remote sensors, in *Proceedings of SPIE Europe Security+Defence*, Berlin, Germany, 2009
42. P.S. Sapaty, Formalizing commander's intent by spatial grasp technology, accepted paper at the International Society of Military Sciences (ISMS) 2012 annual conference, Kingston, Ontario, Canada, 23–24 Oct 2012

43. P. Sapaty, Distributed capability for battlespace dominance, in *Electronic Warfare 2009 Conference & Exhibition*, Novotel London West Hotel & Conference Center, London, 2009
44. P. Sapaty, Global electronic dominance with spatial grasp. Int. J. Commun. Netw. Syst. Sci. **5**(11) (2012)
45. P.S. Sapaty, The world as distributed brain with spatial grasp paradigm, book chapter in *Intelligent Systems for Science and Information*, vol. 542 of the series Studies in Computational Intelligence (2014), pp. 65–85
46. P.S. Sapaty, Global electronic dominance, in *12th International Fighter Symposium*, 6th–8th Nov 2012, Grand Connaught Rooms, London, UK
47. P. Sapaty, Providing global awareness in distributed dynamic environments, in *International Summit ISR*, London, 16–18 Apr 2013
48. P. Sapaty, Ruling distributed dynamic worlds with spatial grasp technology. Tutorial at the international science and information conference 2013 (SAI), London, UK, 7–9 Oct 2013
49. P. Sapaty, M. Sugisaka, J. Filipe, Making sensor networks intelligent, in *Proceedings of the 4th International Conference on Informatics in Control, Automation and Robotics, ICINCO-2007*, Angers, France, 9–12 May 2007
50. P.S. Sapaty, Towards wholeness and integrity of distributed dynamic systems. J. Comput. Sci. Syst. Biol. **9**(3) (2016)
51. P.S. Sapaty, Towards global goal orientation, robustness and integrity of distributed dynamic systems. J. Int. Relat. Diplomacy **4**(6) (2016)
52. P. Sapaty, Gestalt-based integrity of distributed networked systems, in *SPIE Europe Security+Defence*, bcc Berliner Congress Centre, Berlin, Germany, 2009
53. P. Sapaty, M. Sugisaka, M.J. Delgado-Frias, J. Filipe, N. Mirenkov, Intelligent management of distributed dynamic sensor networks. Artif. Life Rob. **12**(1–2), 51–59 (2008)
54. P. Sapaty, Distributed technology for global dominance, in *Proceedings of International Conference Defense Transformation and Net-Centric Systems 2008*, as part of the *SPIE Defense and Security Symposium*, 16–20 Mar 2008, World Center Marriott Resort and Convention Center, Orlando, FL, USA (*Proceedings of SPIE—Volume 6981, Defense Transformation and Net-Centric Systems 2008*, ed. by R. Suresh, 69810T, 3 Apr 2008)
55. P. Sapaty, Distributed technology for global dominance, keynote lecture, in *Proceedings of the Fifth International Conference in Control, Automation and Robotics ICINCO 2008*, *Conference Proceedings*, Funchal, Madeira, Portugal, 11–15 May 2008
56. P.S. Sapaty, M.J. Corbin, P.M. Borst, Mobile WAVE programming as a basis for distributed simulation and control of dynamic open systems. Report at the 4th UK SIWG National Meeting, SGI Reality Centre, Theale, Reading, 11 Oct 1994
57. P. Sapaty, M.J. Corbin, S. Seidensticker, Mobile intelligence in distributed simulations, in *Proceedings of 14th Workshop on Standards for the Interoperability of Distributed Simulations*, IST UCF, Orlando, FL, Mar 1995
58. P.S. Sapaty, P.M. Borst, M.J. Corbin, J. Darling, Towards the intelligent infrastructures for distributed federations, in *Proceedings of 13th Workshop on Standards for the Interoperability of Distributed Simulations*, IST UCF, Orlando, FL, Sept 1995, pp. 351–366
59. P.S. Sapaty, A new technology for integration, simulation, and testing of distributed dynamic systems, in *NATO Proceedings Integration of Simulation with System Testing*, RTO-MP-083, AC/323(SCI-083)TP/43, June 2002, 12 p
60. P.S. Sapaty, M.J. Corbin, P.M. Borst, Towards the development of large-scale distributed simulations, in *Proceedings of 12th Workshop on Standards for the Interoperability of Distributed Simulations*, IST UCF, Orlando, FL, Mar 1995, pp. 199–212
61. P. Sapaty, Distributed interactive simulation and control of collective aerial operations, in *International Conference Military Flight Training*, London, UK, 18–19 Sept 2013
62. P.S. Sapaty, Integration of ISR with advanced command and control for critical mission applications, in *SMi's ISR Conference*, Holiday Inn Regents Park, London, 7–8 Apr 2014
63. P.S. Sapaty, Providing over-operability of advanced ISR systems by a high-level networking technology, in *SMI's Airborne ISR*, Holiday Inn Kensington Forum, London, United Kingdom, 26th–27th Oct 2015

64. P.S. Sapaty, Organization of advanced ISR systems by high-level networking technology. MMC (1) (2016)
65. P. Sapaty, Integral spatial intelligence in ISR applications, in *SMi's ISR Workshop*, Holiday Inn Regents Park, London, 9 Apr 2014
66. P. Sapaty, High-level communication protocol for dynamically networked battlefields, in *Proceedings of International Conference Tactical Communications 2009* (Situational Awareness & Operational Effectiveness in the Last Tactical Mile), One Whitehall Place, Whitehall Suite & Reception, London, UK, 2009
67. P. Sapaty, Tactical communications in advanced systems for asymmetric operations, in *Proceedings of Tactical Communications 2010*, CCT Venues, Canary Wharf, London, UK, 28–30 Apr
68. P. Sapaty, Emerging asymmetric threats, Q&A session, in *Tactical Communications 2010*, CCT Venues, Canary Wharf, London, UK, 28–30 Apr
69. P. Sapaty, Towards massively robotized systems under spatial grasp technology. J. Comput. Sci. Syst. Biol. **9**(1) (2016)
70. P.S. Sapaty, Providing spatial integrity for distributed unmanned systems, in *Proceedings of 6th International Conference in Control, Automation and Robotics ICINCO 2009*, Milan, Italy, 2009
71. P. Sapaty, Unified transition to cooperative unmanned systems under spatial grasp paradigm. Int. J. Trans. Netw. Commun. (TNC) **2**(2) (2014)
72. P. Sapaty, Military robotics: latest trends and spatial grasp solutions. Int. J. Adv. Res. Artif. Intell. **4**(4) (2015)
73. P. Sapaty, From manned to smart unmanned systems: a unified transition, in *SMi's Military Robotics*, Holiday Inn Regents Park, London, 21–22 May 2014
74. P. Sapaty, High-level technology to manage distributed robotized systems, in *Proceedings of Military Robotics 2010*, Jolly St Ermins, London, UK, 25–27 May
75. P. Sapaty, Human-robotic teaming: a compromised solution, in *AUVSI's Unmanned Systems North America 2008*, San Diego, USA, 10–12 June
76. P. Sapaty, M. Sugisaka, Distributed artificial brain for collectively behaving mobile robots, in *Proceedings of Symposium & Exhibition Unmanned Systems 2001*, Baltimore, MD, Jul 31–Aug 2 2001, 18 p
77. P.S. Sapaty, Unified transition to robotized armies with spatial grasp technology, in *International Summit Military Robotics*, London, United Kingdom, 12th Nov–13th Nov 2012
78. P. Sapaty, Towards unified human-robotic societies. Austin J. Robot. Autom. **3**(1) (2017)
79. P. Sapaty, Distributed human terrain operations for solving national and international problems. Int. Relat. Diplomacy **2**(9) (2014)
80. P. Sapaty, M. Sugisaka, R. Finkelstein, J. Delgado-Frias, N. Mirenkov, Emergent societies: an advanced IT support of crisis relief missions, in *Proceedings of Eleventh International Symposium on Artificial Life and Robotics (AROB 11th '06)*, Beppu, Japan, 23–26 Jan 2006. ISBN 4-9902880-0-9
81. P. Sapaty, M. Sugisaka, Advanced networking and robotics for societal engagement and support of elders, in *Proceedings of 16th International Symposium on Artificial Life and Robotics (AROB 16th '11)*, B-Con Plaza, Beppu, Oita, Japan, 27–29 Jan 2011
82. P. Sapaty, distributed missile defence with spatial grasp technology, in *SMi's Military Space*, Holiday Inn Regents Park, London, 4th–5th Mar 2015
83. P.S. Sapaty, Distributed air & missile defense with spatial grasp technology. Intell. Control Autom. Sci. Res. **3**(2), 117–131 (2012)
84. P. Sapaty, Gestalt-based ideology and technology for spatial control of distributed dynamic systems, international gestalt theory congress, in *16th Scientific Convention of the GTA*, University of Osnabrück, Germany, 26–29 Mar 2009
85. P. Sapaty, Crisis management with distributed processing technology. Int. Trans. Syst. Sci. Appl. **1**(1), 81–92 (2006). ISSN 1751-1461
86. P. Sapaty, Night vision under advanced spatial intelligence: a key to battlefield dominance, in *International Summit Night Vision 2013*, London, 4–6 June 2013

87. P. Sapaty, Global management of distributed EW-related systems, in *Proceedings of International Conference Electronic Warfare: Operations & Systems 2007*, Thistle Selfridge, London, UK, Sept 19–20
88. P. Sapaty, A. Morozov, M. Sugisaka, DEW in a network enabled environment, in *Proceedings of the International Conference Directed Energy Weapons 2007*, Le Meridien Piccadilly, London, UK, Feb 28–Mar 1 2007
89. P. Sapaty, High-level organisation and management of directed energy systems, in *Proceedings of Directed Energy Weapons 2010*, CCT, Canary Wharf, London, UK, 25–26 Mar 2010
90. P. Sapaty, Distributed control technology for management of roads with autonomous cars. Int. J. Intell. Unmanned Syst. **5**(2/3) (2017)
91. P. Sapaty, A distributed processing system. European Patent No. 0389655, Publ. 10.11.93, European Patent Office, Munich, 1993
92. P. Sapaty, *Mobile Processing in Distributed and Open Environments* (Wiley, New York, 1999)
93. P. Sapaty, *Ruling Distributed Dynamic Worlds* (Wiley, New York, 2005)
94. P. Sapaty, *Managing Distributed Dynamic Systems with Spatial Grasp Technology* (Springer, Berlin, 2017)

Chapter 2
Spatial Grasp Model and Technology, SGT

2.1 Introduction

The chapter discusses how we are dealing with distributed worlds mentally and how such worlds are organized and managed in physical reality, with the approach offered extrapolating unlimited flexibility of mental world fantasizing, scanning, and perception to higher organizational and management levels of real systems.

The related Spatial Grasp Model (SGM) and resultant Spatial Grasp Technology (SGT), based on spatial patterns self-propagating and self-matching distributed networked worlds, are briefed. Also explained are main principles of the technology's key element—high-level Spatial Grasp Language, SGL—in which all such patterns are expressed. SGL effectively operates with physical, virtual and executive worlds, as well as their any combinations, within the same recursive syntax and semantics.

A general organization of SGL interpreter, its basic data structures and functional units, as well as how the interpretation network operates under spatial scenarios in SGL are explained. The distributed command and control of parallel and fully distributed spatial solutions is organized with the help of history-based distributed dynamic track system, which is also effectively supporting temporary and persistent spatial knowledge and parallel processing and return of remote data.

Much more information on this model, technology, and their implementation is in the related patent [1] and three previous books [2–4]. The history of their development along with previous versions and their networked applications can also be found in many other publications [5–49], so we will not be repeating here the numerous and published details on that matter.

© Springer Nature Switzerland AG 2019
P. S. Sapaty, *Holistic Analysis and Management of Distributed Social Systems*, Studies in Systems, Decision and Control 184, https://doi.org/10.1007/978-3-030-01830-6_2

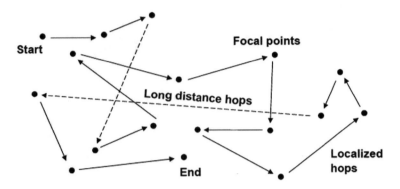

Fig. 2.1 Unrestrained mental movement in a distributed world

2.2 Basics of Spatial Grasp Model

2.2.1 Perception and Handling of Distributed Worlds

2.2.1.1 Mental Activity

Beginning here from how we may be mentally dealing with distributed systems in order to investigate and understand them, also plan certain activities. After concentrating and starting in some focal world point, we can move to other points by following or developing certain scenario in our head and creating and accumulating knowledge fed by our fantasy and previous experience, as in Fig. 2.1, while imagining, even feeling, direct personal presence in this world. The world may have no boundaries for our fantasy, covering any spaces from terrestrial to celestial, where succeeding focal points may be in territorial and conceptual vicinity of the previous points or arbitrarily far away (the latter hops expressed by dashed arrows in Fig. 2.1).

During changing our attention from one focal points to another we may have to go much deeper and more abstractly and fuzzily in thoughts at certain stages with the accumulated scenario and possibly varying intensions and fantasy levels, while subconsciously outlining and investigating new spaces which may be arbitrarily large (shown symbolically in Fig. 2.2 as brainwave-like space coverage). We may be proceeding with this unless define next more or less clear points on which to concentrate further thoughts and investigation, and so on.

2.2.1.2 Physical Activity

If to descend from the extremely flexible and unrestrained mental moving, perception, and reasoning in distributed spaces to the real physical world and activity in it, we may find great difference but also resemblance and match between the two. But

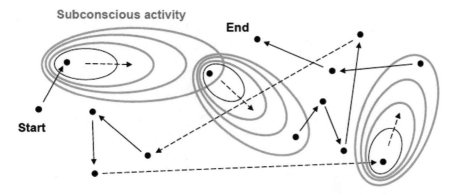

Fig. 2.2 Combination of point-to-point with subconscious activity in finding next points

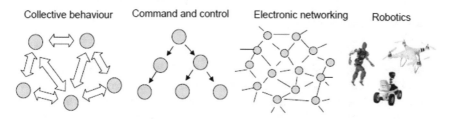

Fig. 2.3 Some real world-related activities

instead of unlimited fantasy and space coverage, we may be having real worlds with limited or even unreachable boundaries, engagement of multitude of humans, electronics, machinery and different means of communication, also sophisticated interactions and control over the resources used. Some typical activities in real world, among many others, and especially related to the current book are shown in Fig. 2.3. Also, in the real world we may have numerous processes taking place and evolving massively, simultaneously, and in parallel, whereas in purely brain activity we are usually proceeding with very limited number of threads at the same time.

2.2.2 Active Distributed Pattern-Based Approach

A high-level distributed vision and processing model has been developed which allows us to describe, understand, simulate, and practically implement vision and activities in any kinds of distributed spaces, from purely mental to fully physical, also any combinations thereof. In this model, the capability of doing practical things in real world can be empowered and enriched by the enormous flexibility and fantasy of our dealing with mental worlds. In most general terms this Spatial Grasp Model (SGM) can be explained as shown in Fig. 2.4.

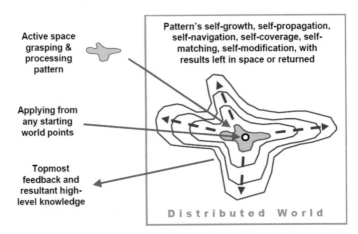

Fig. 2.4 Spatial grasp model main ideas

Any needed observation of and impact on the distributed world is represented as an *active spatial pattern* rather than traditional program, sequential or parallel. This pattern, expressing top semantics and key decisions of the problem to be solved and applied in certain world points, spatially *spreads, grows, covers, and matches* the world at runtime. The pattern can also create, control and change parts of the world or its whole, with final results retained in the distributed space (as passive information or active self-organized systems, generally distributed too) or returned to the application points as detailed data or high level knowledge, whatever required, for a further engagement.

Any number of such patterns and at any time can be applied to same or different points of the world, in any sequence or simultaneously and in parallel, taking also into account that individual patterns can each develop in parallel too. This can provide arbitrary changes and analysis of the world, from its creation to any further modification and processing to the destruction of its parts or the whole, depending on the nature of this world (which may vary from benign to malicious) and the character of tasks formulated on it.

2.2.3 Matching Patterns as Space-Covering Scenarios

Written in a special high-level language (described later in full details) in the form of space covering and processing scenarios, the matching patterns develop in the distributed space using a variety of spatial navigation, processing and control mechanisms, also different types of variables effectively supporting altogether this space navigating and conquering process. All this is empowered by free mobility of the interpreted scenario code that can self-replicate and self-modify during space navigation, also lose utilized parts if not needed any more. As many such details can be

Fig. 2.5 Most general
scheme of space-processing
pattern development

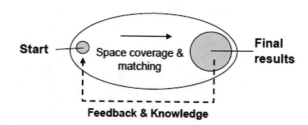

found in the existing publications on this paradigm and model, including the recently
published previous book [4], only very simple examples will be shown below just to
explain the developed model basics.

2.2.3.1   General Scheme of the Pattern's Scenario

The most general scheme of the development of arbitrary complex scenario expressing navigation and search patterns is shown in Fig. 2.5, where the scenario may start
from potentially any but practically most convenient world point. It then spreads and
covers parts of the world or its whole, and the finally obtained results (which may be
multiple and arbitrarily remote from the start) can be left where they were obtained.
The results can also be returned completely or as extracted topmost and particularly
important knowledge to the pattern's application point together with global feedback
control. Both cases for the destinations of final results can also be possible within
the same scenario.

2.2.3.2   Stepwise Space Coverage

Different space navigation scenarios can be combined and integrated into larger scenarios. For example, the constituent scenarios can be developing in distributed space
stepwise, in a forward manner, where each of them starts from all space positions
with obtained results achieved by the previous scenario, like what is expressed in
the scenario language (with local scenarios represented as steps Si) and shown in
Fig. 2.6.

   As SGM widely uses scenario code mobility and self-modification, after application of each step Si its representation is omitted from the scenario body (as not needed
any more). The replicated rest of the scenario will be starting in all positions reached
by Si, with application of the next step Si+1 and transference of the scenario's new
remainder to the new positions reached, and so on, while symbolically expressing
this organization as follows (with character '#' preceding the scenario code transfer
in space):

advance(S1, S2, S3)

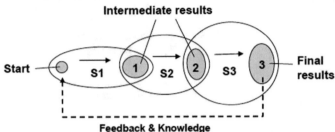

Fig. 2.6 Stepwise space coverage

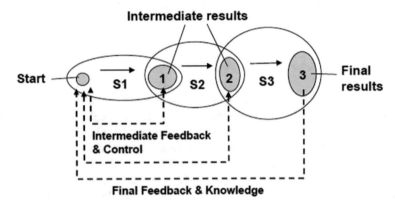

Fig. 2.7 Synchronized forward scenario development

advance(S1, S2, S3) → *Start* → apply(S1) # advance(S2, S3) → *Results1* →

apply(S2) # S3 → *Results2* → apply(S3) → *Results3*

Each new step Si may be immediately starting in any final positions reached by the previous step Si-1 without waiting for other such positions reached (which can be many), i.e. developing fully asynchronously. Possible synchronization of different steps may be useful, depending on the scenario application and nature of scenario steps, and this synchronization can be done in different ways.

For example, this can be achieved with informing the starting position for all intermediate terminations and allowing application of the next steps only after full completion of the previous steps. Such synchronized stepwise development with a feedback to the starting node and subsequent forward control after getting results 1 and 2 is shown in Fig. 2.7, and in the textual form as follows (with symbol 'l' identifying synchronization procedure).

advance(S1, S2, S3) → *Start* → apply(S1) # advance(S2, S3) → *Results1* | →

apply(S2) # S3 → *Results2* | → apply(S3) → *Results3*

Fig. 2.8 Branching space
coverage

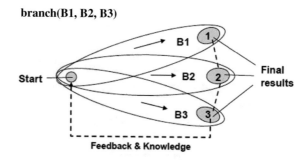

branch(B1, B2, B3)

This combination of feedback to the starting point with forward control can also be used for transferring texts of next steps directly from the starting point each time the results are reached by the preceding step, and not carrying them together with the development of steps themselves, as follows:

advance(S1, S2, S3) → *Start* →

→ apply(S1) → *Results1* |

S2 → | → apply(S2) → *Results2* |

S3 → | → apply(S3) → *Results3*

This latest scheme may have both advantages and disadvantages, depending on remoteness of the intermediate results obtained, complexity of repeated access to them (say, via internal tracks left to them by the evolving steps or directly by their obtained addresses or names), textual length of steps Si, security issues, and other factors.

2.2.3.3 Branching Space Coverage

Under SGM, the space navigation scenario can operate in individual branches (Bi), each originating from the same starting point and producing own positions and results in space, which altogether will represent positions and results on the whole scenario, as linguistically represented and shown in Fig. 2.8.

The development of branching space coverage may be detailed as follows:

branch(B1, B2, B3) → *Start* → apply (B1) → *Results1*

→ apply (B2) → *Results2*

→ apply (B3) → *Results3*

forward(branch(B1, B2), S2)

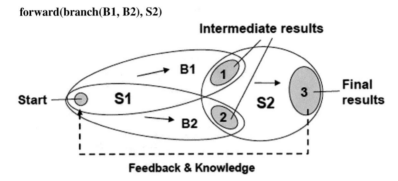

Fig. 2.9 Combination of branching with stepwise coverage

2.2.3.4 Combined Space Coverage

Any composition of stepwise and branching constituent scenarios can be possible for distributed space coverage and matching, as in the example shown in Fig. 2.9, where two branches B1 and B2 represent the first, combined, step S1 after which second step S2 will be developing.

There can be different variants of implementation of this combination.

A *fully asynchronous* case, with replicated scenario S2 transited on forefronts during the development of B1 and B2 and then applied independently in obtained results of B1 and B2, which can be detailed as follows:

advance(branch(B1, B2), S2) → *Start* →

→ apply(B1) # S2 → *Results1* → apply(S2) → *Results31*

→ apply(B2) # S2 → *Results2* → apply(S2) → *Results32*

We may organize *synchronization* after receiving results from B1 and B2, and only then apply S2 in all their positions reached, with the text of S2 being already there, as follows:

advance(branch(B1, B2), S2) → *Start* →

→ apply(B1) # S2 → *Results1* | → apply(S2) → *Results31*

→ apply(B2) # S2 → *Results2* | → apply(S2) → *Results32*

This synchronisation may need additional intermediate feedback and forward control, as shown in Fig. 2.10.

In another development, we may first obtain results by B1 and B2 and only after their full completion transfer scenario S2, delayed in the starting node, to these results (say, effectively using obtained and optimized internal paths to them) considered now as combined ones on S1, and only then deliver and simultaneously apply replicated S2 to these integrated results, as follows.

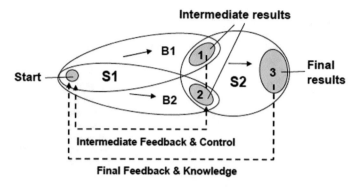

Fig. 2.10 Synchronized branching with stepwise coverage

advance(branch(B1, B2), S2) → *Start* →

→ apply(B1) → *Results1* |

→ apply(B2) → *Results2* | → apply(S2) → *Results3*

S2 → |

A variety of different options on both stepwise and branching space coverage, including mutual dependence, different types of repetition and cycling, echoing remote results, forward and feedback control, and others are contained in the model, which will be explained in the description of its basic language details that follow in this and next chapter.

2.3   Main Features of Spatial Grasp Language, SGL

The Spatial Grasp Language (SGL) is the key element of SGM allowing us to express any matching patterns in a universal, concise, and holistic way with the resulting formal descriptions called *scenarios* rather than *programs* due to nontraditional nature of SGL, also taking into account that these descriptions may be played by humans too.

The language allows us to *directly move through, observe and provide any actions and decisions* in fully distributed environments, whether physical or virtual. In general, *the whole distributed world*, which may be dynamic and active, is considered in SGL as a substitute to traditional computer memory, with multiple "processors" (humans, robots, any manned or unmanned units or devices, etc.) directly operating in it in a cooperative or competitive manner.

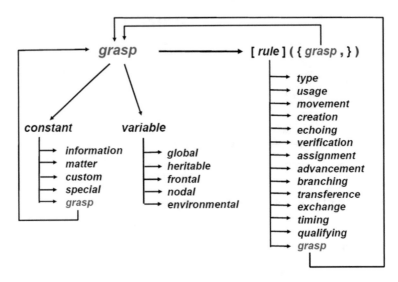

Fig. 2.11 SGL recursive syntax

2.3.1 SGL Recursive Structure

SGL has a recursive structure with its top level shown in Fig. 2.11, which reflects the space-covering-grasping nature of SGM matching patterns (with words in italics representing syntactic categories, square brackets showing an optional construct, braces indicating repetitive parts, and parentheses and comma being the language symbols).

This organization allows us to express any spatial algorithms, create and manage any distributed structures and systems, whether static or dynamic, passive or active, also solve any problem *in*, *on*, and *over* them, and this often can be expressed in a compact and unified way.

An SGL scenario, called *grasp*, applied from a certain world point can just be a *constant* representing the result explicitly or a *variable* containing data assigned to it previously, say, by other SGL scenario (or another branch of the current one) which has already visited this point. It can also be a *rule* expressing some action or definition, optionally supplied with certain parameters (enclosed in parentheses and separated by comma if more than one, which themselves can be any scenarios, i.e. generally represented as *grasp* again).

2.3.2 SGL Basic Elements

Main SGL elements will be presented with some hints on their usage; these being *constants*, *variables* and *rules*.

2.3.2.1   Constants

SGL constants can represent *information*, physical *matter* (physical objects includ-ing), self-identifying *custom* items (relating to information, matter or both), or *special* standard parameters or modifiers used throughout the language in different con-structs. The final option generalizes constant as *grasp* again, potentially allowing it to represent any objects within SGL syntax (passive or with embedded activities) and with any structures, for their further processing by SGL rules.

2.3.2.2   Variables

SGL variables, called "spatial", containing information and/or matter and support-ing different features of distributed scenarios, can be stationary or mobile. They are classified as *global* (with overall common access), *heritable* (event-born and shared by subsequent events), *frontal* (accompanying scenario evolution and propagating in space, i.e. mobile), *nodal* (associated with visited world nodes and locally shared by different processes appearing in them), and *environmental* accessing certain stan-dardized features of external (navigated) and internal (language implementation) environments.

2.3.2.3   Rules

SGL rules, starting their influence in current world positions, can be of different natures and levels—from local matter or information processing to full depth man-agement and control. They can produce results residing in the same or other world positions. The results obtained and world positions reached by rules may become operands and/or starting positions for other rules, with new results and new positions (single or multiple) obtained after their completion, and so on. The rules are cover-ing such language features as *movement, creation, echoing, verification, assignment, advancement, branching, transference, timing, granting, type, usage*, and so on. The final rule's option, *grasp*, provides another level of recursion in SGL where rule names and their compositions may themselves be defined by the results of any SGL scenarios rather than given explicitly.

2.3.3   *More on SGL Details*

The SGL scenario can dynamically *spread* and *process* and *match* the world or its parts needed, with scenario code capable of virtually or physically splitting, repli-cating, modifying, and moving in the distributed spaces being accompanied with transitional data. This movement can take place in single or multiple scenario parts dynamically interlinked under the overall control which is spreading and covering the navigated world too.

2.3.3.1 The SGL Worlds

The language directly operates with:

- *Physical World* (PW), continuous and infinite, where each point can be identified and accessed by physical coordinates expressed in a proper coordinate system (terrestrial or celestial) and with precision given.
- *Virtual World* (VW), which is discrete and consists of nodes and semantic links between them, with both nodes and links capable of containing arbitrary information, of any nature and volume. VW can be considered as finite, but taking into account the rapidly growing world information and the internet growth may potentially be classified as infinite too. It may be hierarchically structured, with higher level nodes containing lower level ones together with links interconnecting them, and links themselves may also contain lower level nodes with their interconnections, and so on.
- *Executive world* (EW) consisting of active "doers" with communication possibilities between them. These may represent any devices or machinery capable of operating on the previous two worlds—including properly equipped humans, robots, mainframes, laptops, smartphones, intelligent sensors, etc. EW can be hierarchical too, with higher level doers (say, groups, organizations, or even societies) consisting of lower level ones down to separate individuals, with proper communications between doer nodes at different levels.

Different kinds of combination of these worlds can also be possible within the same formalism. For example, *Virtual-Physical World* (VPW) may not only be a mere mixture of the two worlds but also their deeper integration where individually named VW nodes can associate with certain PW coordinates and therefore exist in physical reality too. On the other hand, the whole regions of PW (of arbitrary shape and size) may have identifying virtual names, and this naming can be hierarchical. Another possibility is *Virtual-Execution World* (VEW), where doer nodes may have special names assigned to them and semantic relations in between, similarly to pure VW nodes. *Execution-Physical World* (EPW) can have doer nodes pinned to certain PW coordinates as, say, being stationary in these locations; and *Virtual-Execution-Physical World* (VEPW) can combine all features of the previous cases.

2.3.3.2 How SGL Scenarios Evolve

More details on how SGL scenarios self-evolve in distributed environments.

- SGL scenario is considered developing in *steps*, which can be *parallel*, with new steps produced on the basis of previous steps.

- Any step, including the starting one, is always associated with a certain *point* or position of the world (i.e. physical, virtual, executive, or combined) from which the scenario (or its particular part, as there may be many parts working simultaneously) is currently developing.
- Each step provides a resultant *value* (which may be single or multiple, also structured) representing information, matter or both, and a resultant control *state* (as one of possible states ranging by their strength, as shown later). This resultant state may be evaluated and issued in the step's starting point whereas local states can also be issued in the points reached by the step, which may be multiple.
- Different scenario parts may evolve from the same points in *ordered*, *unordered* or *parallel* manner, as independent or interdependent steps-branches.
- Different scenario parts can also spatially *succeed* each other, with new parts evolving from final positions and results produced by the previous parts.
- This potentially parallel and distributed scenario evolution may proceed in *synchronous* or *asynchronous* mode, also their any combinations.
- SGL operations and decisions in evolving scenario parts can use control states and values *returned* from other scenario parts whatever complex and remote they might be, thus combining *forward* and *backward* scenario evolution in distributed spaces.
- Different steps from the same or different scenario parts can be temporarily associated with the same, reached, world points sharing persistent or provisional information in them.
- Staying with world points, it is possible to *change* local parameters in them, whether physical or virtual, thus *impacting* the navigated worlds via these locations.
- Scenarios navigating distributed spaces can *create arbitrary distributed physical or virtual infrastructures* in them, which may operate on their own after becoming active, with or without additional external control. They can also be subsequently (or even during their creation) navigated, updated, and processed by same or other scenarios.
- Overall organization of the world creation, navigation, coverage, modification, analysis, and processing can be provided by a *variety of SGL rules* which may be arbitrarily *nested*.
- The evolving SGL scenario, as already mentioned, can *lose utilized parts* if not needed any more; it can also *self-modify* and *self-replicate* during space navigation, to adjust to unknown environments and optimize communications in distributed systems.

2.3.3.3 Sense and Nature of SGL Rules

Some more light on the general sense and nature of SGL rules which, capable of representing any actions or decisions, may belong to the following categories:

- Elementary arithmetic, string, or logic operation.

- Move or hop in a physical, virtual, execution, or combined space.
- Hierarchical fusion and return of (potentially remote) data.
- Distributed control, sequential and/or parallel, in both breadth and depth of the scenario evolution.
- A variety of special contexts detailing navigation in space, also character and peculiarities of the embraced operations and decisions.
- Type and sense of a value or its chosen usage, for guiding automatic language interpretation.
- Individual or massive creation, modification, or removal of nodes and connecting links in distributed knowledge networks, allowing us to effectively work with arbitrary knowledge structures.
- A rule can be a compound one integrating other rules; it can also be a result of application of another scenario with any complexity and world coverage.

All rules, regardless of their nature, sense or complexity, are pursuing the same unified ideology and organizational scheme, as follows.

- They start from a certain world position, being initially linked to it.
- Perform or control the needed operations in a distributed space, which may be branching, stepwise, parallel and arbitrarily complex, also local and remote.
- Produce or supervise concluding results of the scenario embraced, expressed by control states and values in different points.
- These results reached by the rule's activity may associate with the same (where the rule started) or other world positions, which may be multiple and arbitrarily remote.

This uniformity allows us to effectively compose integral and transparent spatial algorithms of *any complexity and world coverage*, operating altogether under unified and automatic (generally parallel and distributed) control.

2.3.3.4   The Use of SGL Variables

Let us consider some more details on the nature and sense of spatial variables, stationary or mobile, which can be used in fully distributed physical, virtual or executive environments, effectively serving multiple cooperative processes under the unified control. They can be created under the declaration by special rules or by first assignment to them.

- *Global variables*—the most expensive ones, which can serve any SGL scenarios and can be shared by them, also by their different branches. Their locations, mobility capabilities, and life span can depend on the features of distributed environments and SGL implementations. They are recommended to be used in exceptional cases mostly, as other existing types of variables can cover their functionality in many emerging situations in distributed spaces.

- *Heritable variables*—stationary ones, appearing within a scenario step and serving all subsequent, descendent steps, generally multiple and parallel, which can share them in both read and write operations.
- *Frontal variables*—mobile, temporarily associated with the current step and not shared with other parallel steps; they are accompanying scenario evolution, being transferred between subsequent steps. These variables replicate if from a step a number of other steps emerge directly. (The replication procedure, also physical mobility, may have implementation peculiarities if working with physical matter rather than information as frontal variable contents.)
- *Environmental variables*—these allow us to access, analyze, and possibly change different features of physical, virtual and executive words during their navigation, also key parameters of the underlying language implementation system. Most of them are stationary, associated with the world positions reached, but some, especially related to the details of the language interpretation, can be mobile, some even global like those accessing absolute time.
- *Nodal variables*—stationary, being a sole property of the world positions reached by the scenarios. Staying at world nodes, they can be accessed and shared by all activities having reached these nodes under the same scenario identity, and at any time.

These types of variables, especially when used together, allow us to create advanced algorithms working *directly in space*, actually *in between* components of distributed systems rather than *in* them, providing flexible, robust and self-recovering solutions (stealthy too, if needed). Such algorithms can freely self-replicate, partition, spread and migrate in distributed environments (*partially or as an organized whole*), while always preserving overall awareness and global goal orientation.

2.3.3.5 SGL Control States and Their Hierarchical Merge

The following control states can appear after completion of different scenario steps. Indicating local progress or failure they can be used for effective control of multiple distributed processes with proper decisions at different levels.

- Thru—reflects *full success* of the current scenario branch with capability of further development (i.e. indicating successful operation not only *in* but also *through* this step of control). The following scenario steps, if any, will be allowed to proceed from the final location reached by the current step.
- Done—indicates success of the current scenario step with its *planned termination*, after which no further development of this branch from the current step and location reached will be possible. This state can, however, be subsequently changed to *thru* at higher levels by a special rule.
- fail—indicates non-revocable failure of the current branch, with no possibility of further development from the location reached. This state directly relates to the current branch and step only, but can influence decisions at higher levels by special

rules supervising engagement of other branches too (same can be said about the previous two states).

- `fatal`—reports *fatal, terminal failure* with nonlocal effect, triggering massive abortion of all currently evolving scenario processes and removal of the associated temporary data with them, regardless of their current locations and operational success. The scope of this spreading termination may be the whole scenario, by default, or may be restricted by a special containment rule supervising the scenario part within which this state can potentially occur.

These control states appearing in different branches of parallel and distributed scenario at bottom levels can be used to obtain generalized control states at higher levels, up to the whole scenario, in order to make proper decisions for a further scenario evolution. The hierarchical bottom-up merge and generalization of states is based on their comparative importance, or strength, where the stronger state will always dominate while ascending towards the decision root.

For example, merging states `thru` and `done` will result in `thru`, thus generally classifying successful development at a higher scenario level with possibility of further expansion from at least some of its branches. Merging `thru` and `fail` will result in `thru` too, indicating general success with possibility of further development despite some branch (or branches) failed while others remain open to further evolution. Merging `done` and `fail` will result in `done` indicating generally successful termination while ignoring local failures, but without possibility of further development in all these directions.

And `fatal` will always dominate when merging with any other states until its destructive influence is contained within a certain higher level rule, as already mentioned (the latter will itself terminate with *fail* in such a case). So ordering these four states by their powers from maximum to minimum will be as follows: `fatal`, `thru`, `done`, `fail`.

These four states, their merge procedure and the use in control rules are standard, language-embedded features. SGL, as a universal spatial language, also allows us to artificially set up any imaginable control states, with any values and numbers, also any merge or generalization procedures, which may include the mentioned standard ones or be completely different.

2.4   SGL Distributed Interpretation Main Ideas

The SGM, if used in distributed environments, can operate as follows. A network of communicating SGL interpreters (as universal control modules U, see Fig. 2.12) embedded into key system resources (humans, robots, sensors, smart phones, smart watches, etc.) throughout the area of interest collectively interprets high-level mission scenarios written in SGL. Capable of representing any parallel and distributed algorithms these scenarios can start from any node (or nodes), runtime covering the

Fig. 2.12 Distributed interpretation of self-spreading spatial scenarios

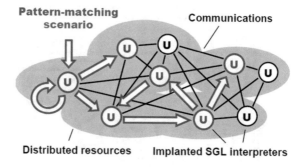

whole world or its parts needed with proper data, operations, and control via the interpretation network.

The self-spreading scenarios can create *knowledge infrastructures*, both passive and active, arbitrarily distributed between system components. Navigated by same or other scenarios, these can effectively support *distributed databases*, *command and control (C2)*, *situation awareness*, and *autonomous decisions*, also simulate *any other* existing or hypothetic computational and/or control models. Many SGL scenarios can simultaneously operate within the same distributed environments, spatially cooperating or competing in the networked space as *overlapping fields of solutions*.

The dynamic network of SGL interpreters covering distributed spaces may have any (including runtime changing) topology and can operate without any central facilities or control, exhibiting at the same time wholeness and high integrity as a system. As will be discussed later, the overall management of distributed evolution of high-level SGL scenarios can be based on a special *track infrastructure* supporting overall awareness, goal orientation, automatic C2, also properly handling various distributed information resources (including their creation, optimization, and cleaning/removal). The distributed execution of SGL scenarios can be effectively implemented in a variety of systems, whether technical or social, with *any types of communications* between their components (electronic, acoustic, visual, postal, even direct voice or paper writing).

SGM and the Spatial Grasp Technology (SGT) resulting from it can convert any collectives (human, robotic, mixed) into holistic systems operating under global goals and capable of acting in complex and unpredictable environments (this may even be effective for collective underwater operations with slow data transfer, due to highly compact operational scenarios in SGL).

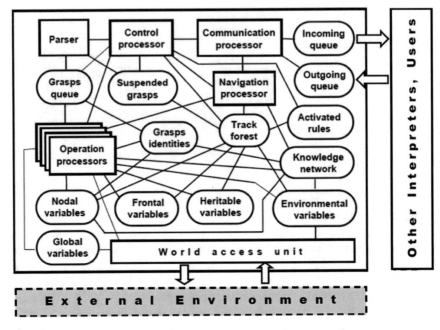

Fig. 2.13 SGL interpreter main components and their interactions

2.5 SGL Interpreter Organization

2.5.1 Components and Structure of SGL Interpreter

The SGL interpreter main components and its general organization are shown in
Fig. 2.13.

The interpreter consists of a number of *specialized functional processors* (shown
by rectangles) working with and sharing specific data structures. These include: Com-
munication Processor (CP), Control Processor (COP), Navigation Processor (NP),
Parser (P), different Operation Processors (OP), and special (external and internal)
World Access Unit (WAU) directly manageable from SGL. Main *data structures* (also
referred to as *stores*) with which these processors operate (shown by ovals) comprise:
Grasps Queue (GQ), Suspended Grasps (SG), Track Forest (TF), Activated Rules
(AR), Knowledge Network (NN), Grasps Identities (GI), Heritable Variables (HV),
Fontal Variables (FV), Nodal Variables (NV), Environmental Variables (EV), Global
Variables (GV), Incoming Queue (IQ), and Outgoing Queue (OQ).

2.5.2 Data Structures of the Interpreter

SGL interpretation network generally serves multiple scenarios or their parallel branches simultaneously navigating the distributed world, which can cooperate or compete with each other. Each interpreter can support and process multiple SGL scenario code which happens to be in its responsibility at different moments of time. This is reflected in the interpreter's basic data structures which hold permanent, persistent, and/or temporary information about the navigated and processed worlds.

2.5.2.1 Grasps Queue

The Grasps Queue (GQ) keeps multiple scenario fragments (syntactically represented as *grasps*) which are parsed and ready for execution in this particular interpreter. Independent from each other and queued by the priority or arrival time, they can be processed sequentially or in parallel (if the latter capabilities exist in the interpreter). Partially executed grasps (say, needing additional execution functionalities) can be returned to the same GQ for further processing.

2.5.2.2 Suspended Grasps

Suspended Grasps (SG) structure keeps deferred parts of SGL scenarios which will be parsed in detail and executed in this interpreter or forwarded to other interpreters when proper conditions are met, control states and/or data availability including. The sequence of recorded grasps in SG *does not represent a queue*, as the time and order of their invocation depend on situations within distributed control of multiple operations in the current and other interpreters.

2.5.2.3 Track Forest

Track Forest (TF) is a special self-optimizing structure reflecting *dynamic history* of spatial evolution of SGL scenarios, allowing us to automatically coordinate and control multiple distributed processes with making decisions at different levels. It preserves *integrity* of the whole set of parallel and distributed processes as a global goal-driven system, also supports existence and controls lifetime of different types of spatial variables described earlier in this chapter and also in the following one. TF spans throughout the navigated world with its interlinked parts kept in different interpreters while forming altogether a *seamless spatial organizational, command and control infrastructure*. Its work will be detailed later in this chapter.

2.5.2.4   Activated Rules

Activated Rules (AR) store represents rules that have been activated and continue working with different scenarios or their branches. Some rules may be waiting for their operands to be completed (say, for obtaining data to be processed or the echoed control states to make decisions). These operands may be arbitrary grasps evolving in the same or in other interpreters and integrated within SGL scenarios by means of track trees in TF. Rules registered in AR are associated with certain TF nodes and the latter, in their turn, with related grasps in GQ and SG, which may represent the rule's operand scenarios currently in active, passive, or suspended states.

The rules in AR may be of any kinds—from simple ones to their aggregates with personal parameters, which may need to be processed first to obtain the values needed. The data-processing rules with all operands completed (i.e. which can be directly executed) are placed into GQ. Rules in AR may form a hierarchy covering via tracks more than one interpreter and operating as an integral system with the help of distributed TF structures.

2.5.2.5   Knowledge Network

The Knowledge Network (KN) store keeps interlinked data on parts of virtual, physical, execution, or combined worlds associated with the current interpreter, this including named or unnamed nodes (which may have physical coordinates) and named links connecting them. These parts may belong to different applications with their specific world domains, which are navigated and processed by different SGL scenarios, where the latter may cooperate, compete, or just ignore each other.

Nodes of KN fragments in the current interpreter may have semantic links with nodes located in other interpreters thus forming integral distributed information infrastructures covering application areas. Any node navigated or created by the current interpreter and belonging to PW, VW, EW or their combinations is immediately registered in the KN store. Except unnamed pure PW nodes, which automatically disappear as soon as all activities related to them cease, all other nodes and relations between them in KN remain regardless of being or not being accessed by SGL scenarios, unless explicitly deleted by special rules.

2.5.2.6   Grasps Identities

The Grasps Identities (GI) are keeping personal colors, or identities, of scenarios that have been currently processed in this interpreter (in an active or suspended way), directing them to the resources related to these scenarios, like nodal variables and KN elements. These can be accessed and shared by different scenarios or their branches with the same identities (or which are aware of these identities while being themselves of other personal colors). This coloring of distributed resources (potentially hierarchical) can be effectively used for protecting own information from unauthorized access while allowing users to work cooperatively with other users on distributed resources from different application areas.

2.5.2.7 Heritable Variables

The Heritable Variables (HV) store keeps all heritable variables with their contents created by SGL scenarios within the current interpreter, these variables being linked to the Track Forest nodes within which they originated. The heritable variables can be accessed by all processes stemming from these track nodes, which may be developing in this or other interpreters. This access is achievable via propagation through the distributed track trees (and between interpreters, if needed) in both read and write mode.

2.5.2.8 Fontal Variables

The Frontal Variables (FV) store holds all frontal variables with their contents registered at the current moment of time within the current interpreter. These variables are linked with the fringe nodes of evolving track trees reflecting space-time evolution of generally hierarchical and parallel processes, being always associated with the latest steps of the interpreted scenarios (having moved to them from the preceding steps). Frontal variables are a sole property of these latest scenario stages; they will be replicated if the scenario splits into branches, becoming a non-sharable property of each new branch. If scenarios have to move to other interpreters, frontal variables will be moving with them too, withdrawing themselves from FV store in the current interpreter.

2.5.2.9 Nodal Variables

Nodal Variables (NV) store keeps all nodal variables with their contents which are linked to nodes in the KN store being their temporary property until the scenarios that created them remain alive in the distributed space (not necessarily in the current interpreter only). For that reason, they are also connected with corresponding nodes in TF store, as after the full completion of scenarios all their track structures, which may spread between interpreters, automatically disappear together with heritable and nodal variables linked to them. Nodal variables are also connected to grasps identities in the GI store through which they can be accessed and shared by all SGL scenarios having these identities (or knowing them).

2.5.2.10   Environmental Variables

Environmental Variables (EV) store keeps special variables with reserved names, which allow us to control and work directly with different space and time features of the worlds created, navigated, and processed, also with internal parameters of the interpreter. Most of these variables instead of having their own contents (like the previous heritable, frontal, and nodal ones) are referring to parameters of other structures or the worlds currently navigated (like those registered in KN) by evolving SGL scenarios. They may also be accessing special hardware or software represented by the World Access Unit (like different timing devices, sensors, and channels), or allowing us to directly communicate with local human users or various external devices. Most of these variables are classified as stationary but some may behave like frontal ones (e.g. scenario colors) temporarily linking with nodes in TF store on their spatial move.

2.5.2.11   Global Variables

Global Variables (GV) store keeps information on global variables, being the most expensive ones, which can be simultaneously used and shared by any scenarios or their different branches, and at any time. Their management is beyond traditional distributed track-supported interpreter organization and can be implemented with the help of Word Access Unit capabilities allowing for direct access of global external stores (or other systems) keeping their contents. In many cases, however, the effect of global variables can be achieved by the use of heritable variables if the latter are declared ahead of the scenario development, but only for individual scenarios, not sets of them where they are independent from each other.

2.5.2.12   Incoming Queue

Incoming Queue (IQ) accepts and stores incoming messages, which may be complete SGL scenarios from users or their parts coming from other interpreters, to be executed in the current interpreter and, possibly, continued in other interpreters afterwards. The messages may also be remotely obtained control states or data to be analyzed and processed in this interpreter or forwarded further. The incoming messages may have different priorities, with control messages being of highest urgency, so the procedures for processing IQ elements must take into account their both arrival order and importance.

2.5.2.13   Outgoing Queue

The Outgoing Queue (OQ) store accumulates obtained results to be passed to the directly connected users, parts of SGL scenarios with accompanying intermediate data to be forwarded to other interpreters for further consideration, control states and data to be returned to other interpreters on their request, as well as direct commands to users or neighboring interpreters. As for the IQ, the procedures of processing OQ messages may combine the first-come-first-served strategy with superiority of command, control and emergency messages exchanged with other parts of the distributed system.

2.5.3 Functional Processors of the Interpreter

Functional Processors carry different SGL interpretation loads, serve interpreter's data structures, and communicate with other interpreters and external users. They also provide system integrity and overall command and control of local and global operations in potentially distributed and dynamic environments.

2.5.3.1 Communication Processor

Communication Processor (CP) receives SGL scenarios or their parts, service and control messages, and requested remote data from the external world or other interpreters, classifying them accordingly and sending to other processors for execution. CP also allows the current interpreter to serve as a transit node in exchanges between other interpreters. The incoming and outgoing messages are optimized by CP for efficient communication with other interpreters, also for processing within the current interpreter. CP directly operates with IQ and OQ, and communicates with Control Processor and Navigation Processor via internal buffers, allowing them to operate asynchronously and in parallel with each other.

2.5.3.2 Parser

Parser (P) carries out syntactic analysis of the scenarios or their parts, extracts control rules in them, decomposes SGL strings into ready for execution elementary grasps (with clarified operands) and the remainders to be suspended for further consideration. The latter can take place when proper conditions are met or values of operands for rules (which themselves may be arbitrary grasps) are finally obtained. Parser also optimizes and compresses SGL code for its further processing (like removing blanks and substituting rule names and special words by short abbreviations). Parser directly operates with GQ and SG, also communicates with Control Processor via internal buffer allowing the two processors to work asynchronously and in parallel with each other.

2.5.3.3   Operation Processors

Operation Processors (OP) unit performs basic analysis and operation procedures over information units and physical matter (or physical objects), expressed by rules. It works directly with GQ, GI, NV, FV, HV, EV, GV, and KN stores. OP also directly communicates with Control Processor, Navigation Processor, and World Access Unit. In case of compound rules, the rest of SGL string taken from GQ for an operation and partially processed in OP can be returned to GQ for a continued execution by other operation processors.

2.5.3.4   Navigation Processor

Navigation Processor (NP) specializes in creation and navigation of network-structured data, operating directly with TF and KN stores. These data networks (more persistent in KN and temporary in TF) may both be distributed between interpreters. In this case NP performs network navigation in the current interpreter while transferring control and orders to NPs in other interpreters when reaching the network boundaries in the current interpreter, for a continued navigation. As regards the history tracks in TF, they may be optimized (by substituting node sets with single nodes when history details become redundant) or removed partially or completely by NP upon termination of SGL scenarios or their parts. NP cooperates directly with Control Processor, Communication Processor, and Operation Processors.

2.5.3.5   Control Processor

Control Processor (COP) provides local and global control and coordination of sequential and parallel processes within the interpreter including classification and forwarding of different types of messages within and between interpreters and decomposition of SGL strings while sending them for execution or suspending till proper conditions met. COP also provides interpretation of all control rules of SGL, supporting the distributed command and control hierarchy based on history tracks, which may spread to and cover other interpreters. COP directly operates on GQ, SG, TF, and AR stores. It also cooperates with CP, P, NP, and OP, supervising collective work of these processors within different SGL interpretation procedures. COP plays the key role in organization and support of track-based distributed management and control.

2.5.3.6   World Access Unit

World Access Unit (WAU) offers an extension to the interpreter's main functionality for interaction with external (and also interpreter's internal) physical and virtual environments, also an interface for integration with other systems like internet, robotic equipment, and all those working with physical matter or objects, global variables too. WAU can be accessed via environmental and global variables as well as directly from processors in OP.

2.5.4   Track-Based Automatic Command and Control

As both *backbone and nerve system* of the distributed interpreter, its hierarchical spatial track system dynamically spans the worlds in which SGL scenarios evolve, providing automatic control of multiple distributed processes. Its part related to the current interpreter is kept in Track Forest store which is interlinked with similar parts in other interpreters, forming altogether *global control coverage*. Self-optimizing in parallel echo processes, this (generally forest-like) distributed track structure provides hierarchical command and control as well as remote data and code access. It also supports spatial variables and merges distributed control states for making decisions at different organizational levels. The track infrastructure can be automatically distributed between different doers during scenario spreading in distributed environments.

2.5.4.1   Track Components and Links with Other Elements

These being as follows:

- *Track nodes* reflecting scenario progress points (or *props*)—the stages through which spatial scenarios evolve and form their development history.
- *Track links* providing transition, succession between consecutive props.
- *Heritable variables* as sole track node properties capable of being accessed by all operations related to the current and subsequent props.
- *Frontal variables* accompanying the scenario evolution and being associated at any time with the latest, fringe track nodes.
- *World nodes* belonging to virtual, physical, executive or combined worlds navigated by the scenario and registered in KN store; these being generally linked with sets of track nodes (as sequences or trees rather than single track nodes).
- *Nodal variables* associated with the world nodes created and/or navigated by SGL scenarios. These variables are also linked with particular track nodes under which they were formed. If these track nodes are removed in the tracks cleaning process caused by termination of scenarios or their parts, these variables will be removed too. Nodal variables will also be deleted if the world nodes to which they belong cease to exist (say, after explicit removal in the scenarios using them).
- *Activated rules* linked with certain track nodes. They start their influence within corresponding props and may use the subsequent track tree (to its full depth) emanating from these nodes for managing and supervising of the rule-related forward and echo operations.

- *Suspended grasps* associated with activated rules and connected to the same track nodes as themselves. These grasps will be subsequently launched by these rules after proper conditions are met, using track infrastructure emanating from this node for their forwarding.

2.5.4.2 Tracks-Based Forward Grasping

In the forward SGL scenario process, the next steps of scenario development can be considered as staying with the same track nodes or forming new track nodes connected to the previous ones by track links. Except reflecting history of scenario evolution, this growing track tree, as already mentioned, is supporting heritable, nodal and frontal variables as well as activated rules and suspended grasps (all being associated with proper tack nodes).

Track nodes are directly associated with particular world nodes at which they appear (following the SGT ideology where all SGL processes are always linked with certain world points where they take place) and this association is inherited by all subsequent track nodes unless they fall into alliance with other world nodes. The latter, in turn, will be inherited by the subsequent track nodes unless shifting to responsibility of other world nodes, and so on.

The track infrastructure can be automatically distributed between different doers during scenario spreading in distributed environments.

2.5.4.3 Echoing via Tracks

After completing the forward stage of SGL scenario discussed above, the track system can return to the starting track node the generalized control state based on termination states in all fringe props, also marking the passed track links with the states returned via them. The states generalization process is based on priority of control states (from strongest to weakest as: *fatal*, *thru*, *done*, and *fail*).

The track system, on the request of higher-level scenario rules, can also collect local data obtained at its fringe props and merge them into a resultant list of values echoed to the starting prop. The track echoing process also optimizes the track system for its further use, for example, by deleting already used and not needed any more items associated with it.

2.5.4.4 Further Forward Development via Tracks

The echo-modified and optimized track system after the previous scenario stage completion can route further grasps to the world positions reached by the previous grasps and defined by fringe track nodes having state *thru*. Heritable variables created in certain track nodes can be accessed from the subsequent nodes in the track system for both reading and writing operations, and at any depth of the evolving track tree, by following links between track nodes in both directions.

More details and examples related to the organization of track system effectively supporting distributed SGL interpretation can be found in the existing publications, and particularly in the preceding book [4].

2.6 Conclusion

We have briefly described the Spatial Grasp Model, Language, and resulting distributed processing and control Technology, SGT, which allows us to effectively work with distributed virtual and physical spaces in parallel and fully distributed mode, without any dedicated central resources. The analysis, processing, and management of distributed systems of any natures is based on self-navigating and self-matching patterns which dynamically create spatial infrastructures throughout distributed worlds in a smog-like, flooding, or even virus-resembling mode, with scenario texts freely moving, self-replicating and self-modifying in distributed environments.

This approach differs fundamentally from any other existing models and technologies by directly working in distributed spaces with the ideological assumption and psychological "feeling" of direct presence in them. This allows us to have extremely compact, fully semantic mission descriptions expressing only main operations and decisions to be undertaken, while often shifting traditional numerous and annoying routines on systems organization and management to the intelligent and networked language interpretation.

SGT may be of particular efficiency for dealing with large social systems and social networks representing them, which will be discussed in the subsequent chapters.

References

1. P. Sapaty, A Distributed Processing System. European Patent No. 0389655, Publ. 10.11.93, European Patent Office Munich, 1993
2. P. Sapaty, *Mobile Processing in Distributed and Open Environments* (Wiley, New York, 1999)
3. P. Sapaty, *Ruling Distributed Dynamic Worlds* (Wiley, New York, 2005)
4. P. Sapaty, *Managing Distributed Dynamic Systems with Spatial Grasp Technology* (Springer, 2017)
5. P.S. Sapaty, in *A Wave Approach to the Languages for Semantic Networks Processing*. Proceedings of International Workshop on Knowledge Representation. Section 1: Artificial Intelligence, Kiev, 1984 (in Russian)
6. P.S. Sapaty, A wave language for parallel processing of semantic networks. Comput. Artif. Intell. **5**(4), 1986
7. S. Varbanov, P.S. Sapaty, in *An Information System Based on the Wave Navigation Techniques*. Abstracts of the International Conference on AIMSA'86, Varna, Bulgaria, 1986
8. P.S. Sapaty, in *The WAVE-0 Language as a Framework of Navigational Structures for Knowledge Bases Using Semantic Networks*. Proceedings of USSR Academy of Sciences. Technical Cybernetics, No. 5, 1986 (in Russian)

9. P.S. Sapaty, I. Kocis, in *A Parallel Network Wave Machine*. Proceedings of 3rd International Workshop PARCELLA'86, Akademie-Verlag, Berlin 1986
10. P.S. Sapaty, in *The Wave Approach to Distributed Processing of Graphs and Networks*. Proceedings of the International Working Conference on Knowledge and Vision Processing Systems, Smolenice, Nov 1986
11. P.S. Sapaty, S. Varbanov, M. Dimitrova, in *Information Systems Based on the Wave Navigation Techniques and their Implementation on Parallel Computers*. Proceedings of the International Working Conference on Knowledge and Vision Processing Systems, Smolenice, Nov 1986
12. Sapaty, S. Varbanov, A. Iljenko, in *The WAVE Model and Architecture for Knowledge Processing*. Proceedings of the Fourth International Conference on Artificial Intelligence and Information-Control Systems of Robots, Smolenice, 1987
13. P.S. Sapaty, The WAVE-1: a new ideology and language of distributed processing on graphs and networks. Comput. Artif. Intell. **5**, 1987
14. P.S. Sapaty, in *WAVE-1: A New Ideology of Parallel Processing on Graphs and Networks*. Proceedings of the International Conference on Frontiers in Computing, Amsterdam, 1987
15. P.S. Sapaty, WAVE-1: a new ideology of parallel processing on graphs and networks. Future Generations Comput. Syst. **4**, 1988 (North-Holland)
16. P.S. Sapaty, The WAVE Model for Advanced Knowledge Processing. Report No. OUEL 1803/89, University of Oxford, England, 1989
17. P.S. Sapaty, The WAVE model for advanced knowledge processing, in *CAD Accelerators*, ed. by A.P. Ambler, P. Agrawal, W.R. Moore (Elsevier Science Publ. B.V., 1990)
18. P.S. Sapaty, in *The WAVE Machine Project*. Proceedings of the IFIP Workshop on Silicon Architectures for Neural Nets, St. Paul de Vence, France, 28–30 Nov, 1990
19. P.S. Sapaty, Logic flow in active data, in *VLSI for Artificial Intelligence and Neural Networks*, ed. by W.R. Moore, J. Delgado-Frias (Plenum Press, New York and London, 1991)
20. P.S. Sapaty, W. Zorn., in *The WAVE Model for Parallel Processing and its Application to Computer Network Management*. International Networking Conference INET'91, Copenhagen, 1991
21. P.S. Sapaty, in *The WAVE Paradigm*. Proceedings of the JICSLP'92 Post-Conference Joint Workshop on Distributed and Parallel Implementations of Logic Programming Systems, Washington, D.C., 13–14 Nov 1992
22. L. Bic, P. Borst, M. Corbin, P. Sapaty, in *The WAVE Control Protocol for Distributed Interactive Simulation*. Proceedings of 11th International Conference on Interoperability of Distributed Simulations, IST UCF, Orlando, FL, 26–30 Sept 1994
23. P.M. Borst, M.J. Corbin, P.S. Sapaty, in *WAVE Processing of Networks and Distributed Simulation*. Proceedings of the HPDC-3 International Conference on San Francisco, IEEE, Aug 1994, pp. 61–69
24. M. Corbin, P.S. Sapaty, in *Using the WAVE Paradigm for Parallel Simulation in Distributed Systems*. Proceedings of the International Conference on ParCo93, Grenoble, France, Sept 1993. *Parallel Computing: Trends and Applications*, ed. by G.R. Joubert, D. Trystram, F.J. Peters, D.J. Evans (North-Holland, 1994)
25. P.S. Sapaty, M. Corbin, P.M. Borst, in *Using the WAVE Paradigm for Modeling and Control of Dynamic Multi-Agent Systems*. Poster at the Artificial Life IV Conference, Massachusetts Institute of Technology, Cambridge, MA, 6–8 July 1994
26. P.S. Sapaty, P.M. Borst, An Overview of the WAVE Language and System for Distributed Processing in Open Networks, Technical Report, Department of Electronic and Electrical Engineering, University of Surrey, June 1994
27. P.S. Sapaty, M. Corbin, P.M. Borst, A. Went, in *WAVE: A New Technology for Intelligent Control in Communication Networks*. Proceedings of International Conference on The Application of RF, Microwave and Millimetre Wave Technologies (M'94), Wembley, UK, Nexus 25–27 Oct 1994, pp. 434–438
28. P.S. Sapaty, M.J. Corbin, P.M. Borst, in Mobile WAVE Programming as a Basis for Distributed Simulation and Control of Dynamic Open Systems. *Report at the 4th UK SIWG National Meeting*, SGI Reality Centre, Theale, Reading, 11 Oct 1994

29. P.S. Sapaty, M.J. Corbin, P.M. Borst, in *Mobile WAVE Programming as a Basis for Distributed Simulation and Control of Dynamic Open Systems*. A Special Session on the WAVE Technology at the 15th International Conference on Distributed Computing Systems, Vancouver, BC, Canada, May–June 1995

30. M.J. Corbin, P.S. Sapaty, Distributed Object-Based Simulation in WAVE. J. Simul. Pract. Theory **3**(3), 157–181 (1995) (Elsevier Science Publishers)

31. F. Merchant, L.F. Bic, P.M. Borst., M.J. Corbin, M. Dillencourt, M. Fukuda, P.S. Sapaty, in *Simulating Autonomous Objects in a Spatial Database Using WAVE*. Proceedings of the 9th European Simulation Multiconference, Prague, June 1995

32. P.S. Sapaty, in *Mobile Wave Technology for Distributed Knowledge Processing in Open Networks*. Proceedings of the Workshop on New Paradigms in Information Visualization and Manipulation, in conjunction with the Fourth International Conference on Information and Knowledge Management (CIKM'95), Baltimore, Maryland, Dec 1995

33. P.S. Sapaty, P.M. Borst, in *WAVE: Mobile Intelligence in Open Networks*. Proceedings of the First Annual Conference on Emerging Technologies and Applications in Communications (etaCOM), Portland, Oregon, IEEE Computer Society Press, May 1996, pp. 192–195

34. P.S. Sapaty, in *WAVE: Creating Dynamic Worlds Based on Mobile Cooperative Agents*. Dartmouth Workshop on Transportable Agents, Dartmouth College, Hanover, New Hampshire, Sep 1996

35. J.C.C. Darling, P.S. Sapaty, in *Distributed Dynamic Virtual Reality in WAVE*. Proceedings of the European Simulation Symposium (ESS-96), Genoa, Italy, Oct 1996, pp. 36–40

36. P.S. Sapaty, *Live Demonstration of the WAVE System and Applications at the Workshop on Mobile Agents and Security 97*, Maryland Center for Telecommunications Research, Department of Computer Science and Electrical Engineering, UMBC, 27–28 Oct 1997

37. P.S. Sapaty, Mobile programming in WAVE, in *Mathematical Machines and Systems*, No. 1, Kiev, January–March 1998, pp. 3–31 (ISSN: 1028-9763)

38. P.S. Sapaty, in *Cooperative Conquest of Distributed Worlds in WAVE*. Proceedings of the Symposium and Exhibition of the Unmanned Systems of the New Millennium, AUVSI'99, Baltimore, MD, July 13–15 1999

39. P.S. Sapaty, Cooperative exploration of distributed worlds in WAVE. Int. J. Artif. Life Robot. (Springer-Verlag, Tokyo) **4**, 109–118 (2000)

40. P.S. Sapaty, in *High-Level Spatial Scenarios in WAVE*. Proceedings of the 5th International Symposium AROB, Oita, Japan, Jan 2000, pp. 301–304

41. P.S. Sapaty, in *Spatial Programming of Distributed Dynamic Worlds in WAVE*. Presentation at the Special Colloquium Internet Challenges, Hasso-Plattner-Institut, University of Potsdam, Berlin, Germany, 4 Oct 2002, 50 p

42. P. Sapaty, M. in *Sugisaka, WAVE-WP (World Processing) Technology*. Proceedings of the First International Conference on Informatics in Control, Automation and Robotics, vol. 1, Setubal, Portugal, 25–28 Aug 2004, pp. 92–102

43. P.S. Sapaty, in *WAVE-WP (World Processing) Technology*. Mathematical Machines and Systems, vol. 3 (2004), pp. 3–17 (ISSN: 1028-9763)

44. P. Sapaty, in *High-Level Technology to Manage Distributed Robotized Systems*. Proceedings of the Military Robotics, Jolly St Ermins, London UK, 25–27 May 2010

45. P. Sapaty, in *Providing Global Awareness in Distributed Dynamic Environments*. International Summit ISR (London, 16–18 Apr 2013)

46. P. Sapaty, in *The World as an Integral Distributed Brain under Spatial Grasp Paradigm*. Book Chapter in Intelligent Systems for Science and Information (Springer, 4 Feb 2014)

47. P. Sapaty, Towards massively robotized systems under spatial grasp technology. J. Comput. Sci. Syst. Biol. **9**(1) (2016)

48. P. Sapaty, A brief introduction to the spatial grasp language (SGL). J. Comput. Sci. Syst. Biol. **9**(2) (2016)

49. P.S. Sapaty, Towards global goal orientation, robustness and integrity of distributed dynamic systems. J. Int. Relat. Diplomacy **4**(6) (June 2016)

Chapter 3
Spatial Grasp Language, SGL

3.1 Introduction

The latest version of SGL will be described particularly suitable for dealing with large distributed social systems and social networks representing them. Extremely compact SGL operational scenarios (often hundred times shorter than in C or Java) are represented as active recursive spatial patterns dynamically matching distributed systems bodies rather than copying the latter and bringing to separated computational resources, as usual. SGL allows us to directly create and process in parallel of large and complex graphs which may be arbitrarily distributed in physical space, with the use of all computational facilities available throughout this space.

All SGL constructs are oriented on operating directly on distributed network bodies, with whole scenario or its parts freely moving through their surfaces and providing proper operations and decisions associated with the network nodes and links, while leaving results in these nodes, carrying them to other nodes or returning to the previous space positions. With similar philosophy and organization, the language allows us operate in distributed physical spaces too, with the resultant physical matter or objects left in space, transferred to other physical locations or returned to the previously visited and process locations.

Also discussed will be possibilities of using some constructs common in other, traditional, languages if this can simplify and shorten the spatial scenarios (within the overall SGL philosophy and its recursive syntax), if such flexibilities are supported by the extended SGL interpreter implementations.

Details on the previous SGL versions and numerous examples of programming in them can be found in the previous books [1–3] as well as in many other publications [4–17].

© Springer Nature Switzerland AG 2019
P. S. Sapaty, *Holistic Analysis and Management of Distributed Social Systems*, Studies in Systems, Decision and Control 184, https://doi.org/10.1007/978-3-030-01830-6_3

3.2 Full SGL Syntax and Main Constructs

We are starting here with the SGL full syntax description, where syntactic categories are shown in italics, vertical bar separates alternatives, parts in braces indicate zero or more repetitions with a delimiter at the right if multiple, and constructs in brackets are optional. The remaining characters and words are the language symbols (including the boldfaced braces).

In the description below, the top syntactic level is separated from the rest of the specification by a double line, whereas lower (same level) parts are separated from each other by single lines.

grasp	→	*constant*	*variable*	*rule* [({ *grasp*, })]

===

constant	→	*information*	*matter*	*custom*	*special*	*grasp*
information	→	*string*	*scenario*	*number*
string	→	` {*character*}'
scenario	→	{ {*character*} }
number	→	[*sign*]{*digit*}[. {*digit*}[e[*sign*]{*digit*}]]
matter	→	" {*character*}"
special	→	thru	done	fail	fatal	infinite	nil	
		any	all	other	allother	current	
		passed	existing	neighbors	direct	
		forward	backward	synchronous	
		asynchronous	virtual	physical	
		executive	engaged	vacant	firstcome

----

variable	→	*global*	*heritable*	*frontal*	*nodal*	*environmental*
global	→	G{*alphameric*}
heritable	→	H{*alphameric*}
frontal	→	F{*alphameric*}
nodal	→	N{*alphameric*}
environmental	→	TYPE	NAME	CONTENT	ADDRESS	
		QUALITIES	WHERE	BACK	PREVIOUS	
		PREDECESSOR	DOER	RESOURCES	LINK	
		DIRECTION	WHEN	TIME	STATE	VALUE	
		IDENTITY	IN	OUT	STATUS

----

rule	→	*type* | *usage* | *movement* | *creation* | *echoing* | *verification* | *assignment* | *advancement* | *branching* | *transference* | *exchange* | *timing* | *qualifying* | *grasp*
type	→	`global`|`heritable`|`frontal`|`nodal`| `environmental`|`matter`|`number`|`string`| `scenario`|`constant`
usage	→	`address`|`coordinate`|`content`|`index`| `time`|`speed`|`name`|`place`|`center`|`range`| `doer`|`node`|`link`|`unit`
movement	→	`hop`|`hopfirst`|`hopforth`|`move`|`shift`| `follow`
creation	→	`create`|`linkup`|`delete`|`unlink`
echoing	→	`state`|`rake`|`order`|`unit`|`unique`|`sum`| `count`|`first`|`last`|`min`|`max`|`random`| `average`|`element`|`sortup`|`sortdown`| `reverse`|`fromto`|`add`|`subtract`|`multiply`| `divide`|`degree`|`separate`|`unite`|`attach`| `append`|`common`|`withdraw`|`increment`| `decrement`|`access`|`invert`|`apply`| `location`
verification	→	`equal`|`nonequal`|`less`|`lessorequal`| `more`|`moreorequal`|`bigger`|`smaller`| `heavier`|`lighter`|`longer`|`shorter`|`empty`| `nonempty`|`belong`|`notbelong`|`intersect`| `notintersect`|`yes`|`no`
assignment	→	`assign`|`assignpeers`
advancement	→	`advance`|`slide`|`repeat`|`align`|`fringe`
branching	→	`branch`|`sequence`|`parallel`|`if`|`or`| `and`|`choose`|`quickest`|`cycle`|`loop`| `sling`|`whirl`|`split`

transference	→	`run	call`
exchange	→	`input	output	send	receive	emit	get`
timing	→	`sleep	allowed`
qualification	→	`contain	release	free	blind	quit	`
		`abort	stay	lift	seize`

3.3   SGL Top Level

General SGL organization is as follows:

grasp ➔ *constant* I *variable* I *rule* [({ *grasp* , })]

From this definition, an SGL scenario, called *grasp*, supposedly applied in some point of the distributed space, can just be a *constant* directly providing the result which will be associated with this point. It can be a *variable* whose content, assigned to it previously when staying in this or other space point (as variables may have non-local meaning and coverage), provides the result in the application point too. It can also be a *rule* (expressing certain action, control, description or context for other rules) optionally accompanied by operands separated from each other by comma (if multiple) and embraced in parentheses. These operands can be of any nature and complexity (including arbitrary and active scenarios themselves) and are defined recursively as *grasps*, i.e. they can be constants, variables or rules with operands (the latter as grasps again), and so on.

Rules, starting in some world point, can organize navigation of the world sequentially, in parallel, or any combinations thereof. They can result in staying in the same application point or can cause movement to other world points with obtained results to be left there, as in rule's final points. Such results can also be collected, processed, and returned to the starting point, the latter serving as the final one on the rule in this case.

The final world points reached after the rule invocation can themselves become starting ones for other rules. The rules, due to recursive language organization, can form arbitrary operational and control hierarchies expressing any sequential, parallel, centralized, localized, mixed and up to fully decentralized and distributed algorithms operating in, with and over large spaces, which can be physical, virtual, or combined.

Let us consider some elementary examples in the SGL syntax.

```
rule1(operand1, operand2, operand3)
```

This may calculate a result from the three given operands (with `rule1` reflecting arithmetic, string or logical operation) and leave it in the same world position where the rule started. The operands may represent directly given constants, variables, or may themselves be arbitrary complex scenarios (covering any regions up to the whole world) with their individual results returned to `rule1` and processed there.

This processing can be stepwise, hierarchical and parallel if operands themselves contain rules which process lower level results, and so on. The final value produced by the rule will be left in the node where the rule started.

The same syntactic example may also have quite different interpretations, depending on what `rule1` and its operands mean.

If, for example, `rule1` is to set moving in a networked virtual world, `operand1` may contain orientation and names of semantic links to be passed while `operand2` provides hints on possible names, contents or addresses of nodes these links should lead to. And `operand3`, for example, can supply additional data on the depth and scale of this propagation in real environment (in case virtual network having recorded relations with the physical world). The resultant world points, from which the following parts of the scenario may be applied, will correspond to the new nodes reached by `rule1`.

If `rule1` is for moving in a two-dimensional physical world, `operand1` and `operand2` may define the two coordinates of the destination node, whereas `operand3` may give the speed of this movement. If propagation is in three-dimensional world, all three operands may define values for different dimensions, and additional `operand4` may happen to be needed for the speed value. The resultant world point for a possible application of other parts of the scenario will correspond to the new physical location reached by the rule.

The operands in the above examples may be constants supplying data directly, variables with previously recorded values, or arbitrary scenarios which may themselves involve complex distributed processes with own rules aimed at providing the needed value (or values, which may be multiple and remote) to `rule1`.

If `rule1` is a control one, it may organize the three operands, as arbitrary scenarios, to evolve and propagate in distributed space in a proper order and from proper starting points.

If it is to organize coordinated stepwise world coverage by the three operand-scenarios, then each subsequent scenario will be applied in parallel from all world positions reached by the previous scenario. The resultant space locations reached under the rule will be determined by all invocations of the *last* scenario (i.e. corresponding to `operand3`). The final results on the whole task will be left in all these final locations after completion of `operand3`, from all or some of which the other scenario parts may be subsequently evolving (using, as their initial data, the final data left there by `rule1`).

If `rule1` is to organize invocation of the three operands independently and possibly in parallel, each from the same rule-starting node, the resultant world positions reached will unite all final positions reached by all scenario operands, with final results left in them, -- to be used by subsequent scenarios starting in all or some of these world points.

By embracing these latest cases with a print-capable `rule2`, we can trigger the return of all final results (i.e. obtained either by all invocations of `operand3` or by all the three operands) to the starting point of `rule2` (i.e. application point of the whole scenario) and exhibit them there.

```
rule2(rule1(operand1, operand2, operand3))
```

Let us consider a bit more complex example, as follows:

```
rule3(rule1(operand1, operand2),
      rule2(operand3, operand4))
```

In this case, imagine that `rule1` and `rule2` organize independent and parallel evolution in space of the scenarios they embrace (i.e. `operand1` and `operand2` in the first case, and `operand3` and `operand4` in the second). And let `rule3` coordinates sequential spatial advances of scenario parts controlled by `rule1` and `rule2`, with `rule2` evolving from all world positions reached by `rile1`.

Using such combinations of these and other SGL rules, also with proper engagement of special spatial variables, as will be shown throughout the book, allows us to express and implement any graph- and network-based spatial patterns which can self-spread and self-match complexly structured distributed networked systems. And, for example, by applying an additional `rule4`, as below, embracing this whole scenario, we may verify the success or failure of this whole distributed pattern-matching operation using history-based SGL distributed coordination and implementation system.

```
rule4(rule3(rule1(operand1, operand2),
            rule2(operand3, operand4)))
```

The latest syntactic example can also be interpreted in a very different way, where the scenario operands coordinated by `rule1`, `rule2`, and `rule3` can be applied to an empty space and create a particular network topology, with context-like `rule4` supplying the whole scenario with the needed creative power. In general cases, similar SGL-based active patterns can combine propagation through existing spaces with their extension, modification, creation, and arbitrary processing.

3.4 SGL Constants

The definition of constants, which can be of different types, is as follows:

constant ➔ *information | matter | custom | special | grasp*

3.4.1 Information

Information constants can be of the following categories:

information ➔ *string | scenario | number | custom*

3.4.1.1 String

A string can be represented in most general way as a sequence of characters embraced by opening-closing single quotation marks:

string ➜ `'{character}'`

This sequence should not contain other single quotes inside unless they appear in opening-closing pairs, with the nesting allowed to any depth.

Examples: `'John'`,   `'Peter and Paul'`,   `'Bob is said to 'sometimes trust' Mike'`.

If single words representing information are not intersecting with other language constructs, the quotes around them can be omitted, as follows:

`John, Peter, Paul, Bob, Mike`

3.4.1.2   Scenario

Another string representation may be in the form of explicit SGL scenario body:

scenario ➜ **{** `{character}` **}**

Instead of single quotes, a sequence of characters can be placed into opening-closing curly brackets, or braces **{ }**, shown here in bold, which can be used inside the string and nested in pairs too. Braces will indicate the text as a potential *scenario* code which can be immediately optimized (like removing unnecessary spaces and/or adjusting to the standard SGL syntax after using constructs typical to other programming languages for convenience, as mentioned later. If single quotes are used to embrace texts as SGL code, such code optimization will have to be done each time *during* its interpretation, and not before, with the original text remaining intact.

3.4.1.3   Number

Numbers can be represented in a standard way, similar to traditional programming languages, generally in the form:

$$[sign]\{digit\}[.\{digit\}[E[sign]\{digit\}]]$$

(with brackets identifying optional and braces repeating characters).
    Examples:
`105, 88.56, -15, 3.3E-5`

Numbers can also use words instead of digits and accompanying characters like sign and dot (with underscore as separator if more then one word needed to represent them). The four examples shown above may also look like:

(a)   with mixed representation:

```
hundred_five, eighty_eight.56, minus_fifteen,
three.3E-five
```

(b)   up to the full wording:

```
one_zero_five, eight_eight_dot_fifty_six,
minus_one_five, three_dot_three_E_minus_five
```

3.4.2   Physical Matter

Physical *matter* (physical objects including) in most general way can be reflected in SGL by a sequence of characters embraced by opening-closing double quotation marks:

matter ➜ "{*character*}"

Examples:

"truck", "white sand", "brick", "water"

The above mentioned self-identifying constants (like strings, scenarios, numbers, and matter) can also be set up by explicitly naming their types with the use of corresponding rules, as shown later.

3.4.3   Custom Constants

Other self-identifiable, *custom*, constants can be incorporated into SGL too, if do not conflict with the language syntax, to be directly processed by updated SGL interpreter. They can represent standing without quotes or braces information and physical matter different substances or objects, as well as their combinations.

For example, these may be coordinates in physical spaces like:

x17.5, y44.2, z-77 or their integration:

X17.5_y44.2_z-77

They may be internet addresses, for example:

www.amazon.com

Or combined physical-information-executive cases like:

sand_7_ton, soldier_Peter, robot_Smart

Additional type-clarifying rules can also be introduced for defining different kinds of new constants, which may now have any textual representations.

3.4.4   Special Constants

Special reserved constants (as free standing words, without quotes) may be used as standard parameters (or modifiers) in different language rules, with most frequently used ones listed as follows.

special ➜ thru | done | fail | fatal | infinite | nil | any | all |
other | allother | current | passed | existing | neighbors | direct
| forward | backward | neutral | synchronous | asynchronous |
virtual | physical | executive | engaged | vacant | firstcome |
unique

We will be explaining here their possible meanings and applications.

thru—indicates (or artificially sets up) control state of the scenario in the current world point as an *absolute success* with possibility of further scenario evolution from this particular point. This does not influence scenario developments in other world points.

done—indicates (or sets up artificially) control state as a *successful scenario termination* in the current world point, with blocking further scenario development from this particular point. Not influencing scenario developments in other world points.

fail—indicates (or artificially sets up) scenario control state as *failure* in the current world point, without possibility of further scenario development from this particular point. Not influencing scenario developments in other world points.

fatal—indicates (or sets artificially) control state as nonlocal *fatal failure* starting in the current world point and causing massive removal of all active distributed processes with related local data in this and other world points reached by the same scenario (which may have independent and parallel branches). The destructive influence of this state may be contained at some levels by special rules explained later.

infinite—indicates infinitely large value.

nil—indicates no value at all.

any, all, other, allother—stating that any one chosen, all (the current one including), any other, or all other (the current one excluding in the last two cvases) elements under consideration can be used by some rule.

current—refers to the current element (like node) only, for example, for its further consideration or reentering (possibly, with proper conditions).

passed—informing that the mentioned elements (like world nodes) have already been passed by the current scenario branch on the way to the current point (and can possibly be accessed easier by using SGL history-based distributed control).

existing—hinting that world nodes with given names which are under consideration already exist and should not be created again (i.e. duplicated).

neighbors—stating that the nodes to be accessed are among direct neighbors of the current node, i.e. located within a single hop from it by the existing links.

direct—stating that the mentioned nodes should be accessed or created from the current node directly, without consideration of possible direct semantic links to them (even if such links happen to exist in the case of accessing already existing nodes).

forward, **backward**, **neutral**—allowing us to move from the current node via existing links along, against or regardless their orientations (ignored when dealing with non-oriented links, which can always be traversed in both directions).

synchronous, **asynchronous**—a modifier setting synchronous or asynchronous mode of operations induced by different rules.

virtual, **physical**, **executive**—indicating or setting the type of a node the scenario is currently dealing with (the node can also be of a combined type, having more than one such indicator, with maximum three).

engaged, **vacant**—indicating or setting the state of a resource the current scenario is dealing with (like, say, human, robot, or any other physical, virtual or combined world node).

firstcome—allows the current scenario with its unique identity to enter the world nodes only first time (the capability based on internal language node marking mechanisms used, for example, in effective blocking of unplanned or unwanted cycling).

3.4.5 Compound Constants, or Grasps

Constants can also be compound ones, using the recursive *grasp* definition of SGL syntax to represent nested, hierarchical structures consisting of multiple (elementary or again compound) constants, which particularly for constants can be expressed as follows:

constant ➔ *rule* ({ *constant* , })

Different SGL rules explained later may be used for such structuring, while others may added for particular applications. Any SGL scenario with all its rules and other constructs can also be considered, if needed, as a *structured constant* to be analyzed and modified by the existing SGL rules.

3.5 SGL Variables

There are five types of SGL variables, called *spatial*, serving quite differently multiple cooperative processes in distributed virtual, physical, executive and combined spaces, as follows:

variable ➔ *global* | *heritable* | *frontal* | *nodal* | *environmental*

Different types of variables can be self-identifiable by the way their names are written. Their names can also have any representations if explicitly declared by special rules explained later. The meanings of these variables and details of their usage are explained in the following sections.

3.5.1 Global Variables

This is the most expensive type of SGL variables with their names starting with capital G and followed by arbitrary sequences of alphabetic letters and/or digits:

global ➔ G{*alphameric*}

These variables can exist only in single copies with particular names, being common for both read and write operations to all processes of the same scenario, regardless of their physical or virtual distribution and world points they may be associated with at different moments of time. Global variables can be created by first assignment to them within any scenario branch and used by the whole scenario. They cease to exist when the scenario that created them terminates.

Examples:

```
Globe, Genesis12, Globaltechnology
```

3.5.2 Heritable Variables

The names of these variables should start with capital H if not defined by a special rule:

heritable ➔ H{*alphameric*}

Heritable variables, being created by first assignment to them at any scenario stage in a related world point, are becoming common for read-write operations *for all subsequent* scenario operations (generally multiple, parallel and distributed) evolving from this particular point. This means that these variables are unique only within the concrete hereditary scenario developments, to all their depth, but can be independently created and used, with same names including, in other process inheritances.

The life time of these variables depends on the continuing activity of processes that can potentially inherit them, with their automatic removed after all such processes terminate.

Examples:

```
H214b, Highlevel, Huge22
```

Heritable variables can also model global variables if declared at the very beginning of the scenario, as all scenario developments will be using and sharing them afterwards. But global and heritable variables may have different implementations where each of them may have advantages under certain conditions.

3.5.3   Frontal Variables

These are mobile type variables, their names starting with capital F, which are propagating in distributed spaces with their contents always being on forefronts of the
evolving scenarios:

frontal ➜ F{*alphameric*}

Each of these variables is serving only the current scenario branch operating in
the current world point, and cannot be shared with other braches evolving in the
same or other world points, always moving together with scenario control. If the
scenario splits into individual branches in the same world point or when moving to
other points, these variables are replicated with the same names and contents and
serve these branches independently.

Depending on applications, there may be different variants of dealing with frontal
variables holding physical matter or physical objects as their contents, rather than
information, especially in relation to their physical movement and automatic replication in distributed environments.

Examples:

```
Frontal5, Freeworld245, Final
```

3.5.4   Nodal Variables

Variables of this type, with identifiers starting with capital N, are a temporary and
exclusive property of the world points visited by SGL scenarios.

nodal ➜ N{*alphameric*}

Capable of being shared by all scenario branches visiting these nodes, they are
created by first assignment to them while staying in the node and reside there until
the whole scenario remains active, being automatically removed after the scenario
termination. These variables are also deleted when nodes they associate with are
removed by same or other scenarios.

Examples:

```
New, Next05, Neveragain
```

3.5.5   Environmental Variables

These are special variables with reserved names which allow us to have access to
physical, virtual, and execution worlds when they are navigated by SGL scenarios,
also to some important internal parameters of the language interpretation system
itself.

environmental ➜ TYPE I NAME I CONTENT I ADDRESS I QUALITIES I WHERE I BACK I PREVIOUS I PREDECESSOR I DOER I RESOURCES I LINK I DIRECTION I WHEN I TIME I STATE I VALUE I IDENTITY I IN I OUT I STATUS

These variables have specific names, all written in capital letters, with brief explanation of their sense and usage following.

TYPE—indicates the type of a node the current scenario step associates with. This variable returns the node's type (i.e. virtual, physical, executive, or their combination as a list with more than one value). It can also change the existing node's type by assigning to it another value (simple or combined).

NAME—returns name of the current node as a string of characters (only if the node has virtual or executive dimension or both). Assigning to this variable when staying in the node can change the node's name.

CONTENT—returns content of the current node (if it has virtual or executive dimension, or both) as arbitrary constant (say, any text in quotes, vector or nested structure consisting of multiple texts, etc.) if this content had been assigned to this node previously, when staying in it. Assigning to this variable when staying in the node can change the node's content. In the case of executive nodes (like human, robot, server, etc.), CONTENT may return, if allowed, some existing specific data like dossier on a human or technical characteristics for a robot and may be read-only.

ADDRESS—returns a unique address of the current virtual node (or node with virtual dimension). This is a read-only variable as node addresses are set up automatically by the underlying distributed interpretation system during creation of virtual nodes, or by an external system (for example, this can be an internet address of the node). The returned address can be remembered and used afterwards for direct hops to this node, form any other positions of the distributed virtual world.

QUALITIES—identifies a list of selected formalized physical parameters associated with the current physical position, or node, depending on the chosen implementation and application (for example, these may be temperature, humidity, air pressure, visibility, radiation, noise or pollution level, density, salinity, etc.). These parameters (generally as a list of values) can be obtained by reading the variable. They may also be attempted to be changed (depending on their nature and implementation system capabilities) by assigning new values to QUALITIES, thus locally influencing the world from its particular point.

WHERE—keeps world coordinates of the current physical node in the chosen coordinate system (the node can be a combined one also having virtual and/or executive features). These coordinates can be obtained by reading this variable. Assigning a new value to this variable (with possible speed added) can cause physical movement into the new position (with same node's identity, all its information surrounding, and control and data links with other nodes).

BACK—keeps internal system link to the preceding world node (virtual, executive or combined one) allowing the scenario to most efficiently return to the previously occupied node, if needed. This variable exclusively refers to internal interpretation mechanisms (its content cannot be lifted, recorded, or changed from the scenario level) and can be used in direct hop operations only.

PREVIOUS—refers to the absolute and unique address of the previous virtual node (or combined one with executive and/or physical dimensions), allowing us to return to the node directly. This return may be on a higher level and therefore more expensive than using BACK, but the content of PREVIOUS, unlike BACK, can be lifted, recorded, and used elsewhere in the scenario (but not changed, similar to ADDRESS).

PREDECESSOR—refers to the name of preceding world node (the one with virtual or executive dimension, visited just before the current one). Its content can be lifted, recorded and subsequently used, for organization of direct hops to this node too (on highest and most expensive level, however). Assigning to PREDECESSOR in the current node can change the name of the previous node.

DOER—keeps the name of the device (say, laptop, robot, smart sensor, or a specially equipped human) which interprets the current SGL code in the current world position. This device can be initially chosen for the scenario automatically from the list of recommended devices or just picked up from those expected available. It can also be appointed explicitly by assigning its name to DOER, causing the remaining SGL code to move immediately into this device and execute there (the change of the device can also be done automatically by the distributed SGL interpreter).

RESOURCES—keeps a list of available or recommended resources (human, robotic, electronic, mechanical etc., by their types or names) which can be used for planning and execution of the current and subsequent parts of the SGL scenario. This list can also contain potential doers which may appear (by their names) in variables DOER. The contents of RESOURCES can be changed by assignment, and in case of distributed SGL interpretation and spatial branching they may be replicated or partitioned by the distributed SGL planning and optimization mechanisms, to properly serve different branches using RESOURCES associated with them.

LINK—keeps the name (same as content) of the virtual link which has just been passed. Assigning a new value to it can change the link's content/name. Assigning nil or empty to LINK removes the link passed.

DIRECTION—keeps direction (along, against, or neutral) of the passed virtual link. Assigning to this variable values like plus, minus, or nil (same as +, −, or empty) can change its orientation or make the link non-oriented.

WHEN—assigning value to this variable sets up an absolute starting time for the following scenario branch (i.e. starting with the next operation), thus allowing us to suspend and schedule operations and their groups in time.

TIME—returns current absolute system time, as read-only global variable.

STATE—can be used for explicit setting resultant control state of the current scenario step by assigning to it one of the following constants: thru, done, fail, or fatal, which will influence further scenario development from the current world point. (These states are also generated implicitly and automatically on the results of success or failure of different operations, belonging to the internal control mechanisms of SGL scenarios.) Reading STATE will always return thru as this could be possible only if the previous operation terminated with thru too, thus letting this operation to proceed. A certain state explicitly set up in this variable can also be used at higher levels (possibly, together with termination states of other branches) within distributed control provided by nested SGL rules, whereas assigning `fatal` to STATE may cause abortion of multiple distributed processes with associated data.

VALUE—when accessed, returns the resultant value of the latest operation (say, an assignment to a variable, unassigned result of arithmetic or string operation, or just naming a variable or constant). Such implicit or explicit (by special rule) assignment to VALUE leaves its content available to the next operation.

IDENTITY—keeps identity, or color, of the current SGL scenario or its branch, which propagates together with the scenario and influences grouping of different nodal variables under this identity at world nodes. This allows different scenarios or their branches with personal identities to be protected from influencing each other, even if they are using same named nodal variables. However, scenarios with different identities can penetrate into each other information fields if they know the other's colors, by temporarily assigning the needed new identity to IDENTITY (to perform cooperative or stealth operations) while restoring the previous color afterwards. Any numerical or string value can be explicitly assigned to IDENTITY. By default, different scenarios are keeping the same value in IDENTITY assigned automatically at the start, thus being capable of sharing all information at navigated nodes, unless change their personal color themselves.

IN—special variable requesting and reading data from the outside world in its current point. The received data is becoming the resultant value of the reading operation.

OUT—special variable allowing us to issue information from the scenario to the outside world in its current point, by assigning the output value to this variable.

STATUS—retrieving or setting the status of (especially doer) node in which the scenario is currently staying (like `engaged` or `vacant`, possibly, with a numerical estimate of the level of engagement or vacancy). This feedback from implementation layer can be useful for a higher-level supervision, planning and distribution of resources executing the scenario rather than doing this fully implicitly.

Other environmental variables for extended applications can be introduced and identified by unique words in all capitals too, or they may use any names if explicitly defined by special rules, as shown later. As can be seen, most environmental variables are behaving as stationary ones, except RESOURCES and IDENTITY, which are mobile. The global variable TIME may be considered as stationary too, but can also be implemented in the form of individual TIME clocks regularly updating their system time copies and propagating with scenarios as frontal variables.

3.6   SGL Rules

The main types of SGL constructs called *rules* are as follows:

rule ➜ *type* | *usage* | *movement* | *creation* | *echoing* | *verification* | *assignment* | *advancement* | *branching* | *transference* | *exchange* | *timing* | *qualifying* | *grasp*

The concept of *rule* is dominant in SGL not only for diverse activities on data, knowledge and physical matter, but also for overall management and control. This provides integral and unified approach for expressing everything that might take place or even come to mind in large dynamic worlds, and generally in parallel and distributed mode. This section describes the main repertoire of SGL rules with summaries of their features.

3.6.1   Type

These rules explicitly assign types to different constructs, with their existing repertoire following.

type ➜ global | heritable | frontal | nodal | environmental | matter | number | string | scenario | constant

global, **heritable**, **frontal**, **nodal**, **environmental**—allow different types of variables to have any alphanumeric names rather than those oriented on self-identification, as explained before. These names will represent variables with needed types in the subsequent scenario developments unless redefined by these rules. As regards environmental variables, their names differing from the standard ones (also new kinds of such variables) will need special adjustment on the implementation layer.

matter, **number**, **string**, **scenario**, **constant**—allow arbitrary results obtained by the embraced scenario, with any their current types, to properly represent the needed values rather than using self-identifiable representations mentioned before (with automatic internal type adjustments, conversions, and optimizations if necessary).

3.6.2   Usage

These rules explain how to use the information units they embrace, with main variants as follows:

usage ➜ address | coordinate | content | index | time | speed | name | center | range | doer | node | link

They are adding certain flexibility to representation of SGL scenarios where strict order of operands in rules and presence of them all may not be absolute.

address—identifies the embraced value (which may also be an arbitrary scenario producing this value or values, if multiple) as an address of a virtual node.

coordinate—identifies the embraced value as physical coordinates (say, one, two, or three dimensional), and may also be a list of coordinates given directly or represented by a result of application of an arbitrary scenario.

content—identifies the embraced operand as a content (or contents) which may, for example, relate to identical values of elements in a list.

index—identifies the embraced operand as an index (or indices), which may represent orders of values of elements in a list.

time—tells that the embraced operand represents time value.

speed—tells that the embraced operand represents a value of speed.

name—identifies the embraced operand as a name (say, of a virtual or executive node or nodes).

center—depending on applications, indicates that virtual address or physical coordinates embraced may relate to the center of some world region.

range—identifies virtual or physical distance that can, for example, be used as a threshold for certain operations in distributed spaces, especially those evolving from a chosen center.

doer—identifies the embraced name or any other value as belonging to executive node (like human, robot, server, smart phone, etc.).

node—(or **nodes**, if more appropriate) identifies the embraced value or values as keeping names of nodes having virtual or/and executive dimensions.

link—(or **links**, if more appropriate) informs that the embraced value or values represent names of links connecting virtual nodes in knowledge networks.

3.6.3 Movement

The movement rules have the following options:

movement ➜ hop | hopfirst | hopforth | move | shift | follow

They may result in virtual hopping to the existing nodes (the ones having virtual or/and executive dimensions) or in real movement to new physical locations, subsequently starting the remaining scenario (with current frontal variables and control) in the nodes reached. The resultant values of such movements are represented by names of reached nodes (in case of virtual, executive, or combined nodes) or nil in case of pure physical nodes, with control state thru in them if the movement was successful. If no destinations have been reached, the movement results with state fail and value nil in the rule's starting node. These rules have the following options.

hop—sets electronic propagation to node(s) in virtual, execution, or combined spaces (the latter may have physical dimension too), directly or via semantic links connecting them with the starting node. In case of a direct hop, except destination node name or address, special modifier direct may be included into parameters of the rule. If the hop is to take place from a node to a particular node via existing link, both destination node name/address and link name (with orientation if appropriate) should be among parameters of the rule.

This rule can also cause independent and parallel propagation to a number of nodes if there are more than one node connected to the current one by same named links, and only link name mentioned (or given by indicator `all`, for all links involved). In a more general case, parallel hops can be organized from the current node if the rule's parameters are given by a list of possible names/addresses of destination nodes and a list of names of links which may lead to them (`direct` and/or `all` indicators can be used here too).

hopfirst—modification of the `hop` rule allowing it to come to a node only first time (for the scenario with certain identity), which is based on internal language interpretation mechanism properly marking the nodes visited. The use of this rule can be equivalent to the previous rule `hop` with modifier `firstcome` (which can also be used in other cases, like new linking to the existing nodes, as mentioned later).

hopforth—modification of the `hop` rule allowing it to hop to a node which is not the one just visited before, i.e. excluding the return to the previous node. It may be considered as a restricted variant of the previous rule. Both rules can be useful for effective blocking of looping in networked structures, if useful for certain scenarios.

move—sets real movement in physical world to a particular location given by coordinates in a chosen coordinate system. The destination location becomes a new temporary node with no (or `nil`) name, which disappears when all current scenario activities leave it for other nodes. The location reached may, however, become a permanent node if also assigned a virtual dimension (possibly, virtual name too), after which such combined node can be found and entered by other scenario branches. Speed value for such movement may be used as an additional parameter.

shift—differs from the `move` only in that movement in physical world is set by deviations of physical coordinates from the current position rather than by their absolute values.

follow—allows us to move in virtual or physical spaces using already (internally) recorded and saved paths to the destinations needed, to enter them second time, as will be explained later.

3.6.4   Creation

These rules have the following options:

creation ➔ `create | linkup | delete | unlink`

They create or remove nodes and/or links leading to them during distributed world navigation. After termination of the creation rules, their resultant values will correspond to the names of reached nodes with termination states `thru` in them, and the next scenario steps if any will start from all these nodes. After removal of the destination nodes and/or links leading to them, the resultant world position will be the rule's starting node with the same value as before and control state `thru`. If the creation or removal operation fails, its resultant value will be `nil` and control state `fail` in the node the rule started, thus blocking any further scenario development from this node.

create—starting in the current world position, creates either new virtual link-node pairs or new isolated nodes. For the first case, the rule is supplied with names and orientations of new links and names of new nodes these links should lead to, which may be multiple. For the second case, the rule has to use modifier `direct` indicating direct nodes creation. If to use modifiers `existing` or `passed` for the link-node creation hinting that such nodes already exist or, moreover, have already been passed by this scenario, only links will be created to them by `create`. Same will take place if nodes are given by their addresses, the latter always indicating their existence. The modifier `firstcome`, if used, will not allow entering same nodes more than once by the same colored scenario.

linkup—restricts the previous rule by creating only links with proper names from the current node to the already existing nodes given by their names or addresses. Using modifier `passed` here may also help it to narrow direct search of the already existing nodes. Also, the modifier `firstcome`, if used, will not allow entering same nodes more than once, as before.

delete—removes links together with nodes they should lead to, starting from the current node. Links and nodes to be removed should be either explicitly named or represented by modifiers `any` or `all`. Using modifier `direct` instead of link name together with node name will allow us to remove such node (or nodes) from the current node directly. In all cases, when a node is deleted, its all links with other nodes will be removed too.

unlink—removes only links leading to neighboring nodes where, similar to the previous case, they should be explicitly named or modifiers `any` or `all` used instead.

The above mentioned creation rules, depending on implementations, can also be used in a broader sense and scale, as *contexts embracing arbitrary scenarios* and influencing hop operations within their scope, i.e. the same scenarios will be capable of operating in creation and deletion modes too, not only for navigating the existing networks. These contexts can influence both links and nodes when dealing with existing networks or empty spaces in which such networks should be created.

3.6.5 Echoing

This class of rules, oriented on various aspects of data and knowledge processing, contains the following rules which may use local and remote values for different operations:

echoing ➜ state|rake|order|unit|unique|sum|count|first|last |min|max|random|average|element|sortup|sortdown|reverse | fromto | add | subtract |multiply| divide | degree | separate | unite|attach|append| common|withdraw| incerent |decrement| access| invert|apply| location

The listed rules use terminal world positions, their control states, and associated final values (which may be local or arbitrarily remote) from the embraced scenario to obtain the resultant state and value in the location where the rule started. This location will represent the rule's single terminal point from which the rest of the scenario, if any, can develop further. The usual resultant control state for these rules is `thru` (state `fail` occurs only if certain terminal values happen to be unavailable or the result is unachievable, say, as division by zero). Depending on the rule's semantics, the resultant value may happen to be compound, like a list of values, which may also be hierarchically nested.

state—returns the resultant generalized state of the embraced SGL scenario upon its completion, whatever its complexity and space coverage may be. This state being the result of ascending fringe-to-root generalization of terminal states of the scenario embraced, where states with higher power (their sequence from maximum to minimum as: `fatal`, `thru`, `done`, `fail`) dominate in this potentially distributed and parallel process. The resultant state returned is treated as the *resultant value* on the rule, the latter always terminating with own final control state `thru`, even in the case of resultant `fatal`.

rake—returns a list of final values of the scenario embraced in an arbitrary order. This order may, for example, depend on the order of completion of branches and/or times of reaching their final destinations. Additionally using `unique` as modifier, described before, will result in collecting only unique values, i.e. with possible duplicates omitted/removed.

order—returns an ordered list of final values of the scenario embraced corresponding to the order of launching related branches rather than the order of their completion. For potentially parallel branches, these orders may, for example, relate to how they were activated, possibly, with the use of time stamps upon invocation. Similar to the previous rule, modifier `unique` can be used too for avoiding duplicate values.

unit—returns a list of values while explicitly reminding that this list represents an integral unit which should not be mixed with other elements possibly returned from other final points and which may be represented as integral units too, to form the results as a list of lists, which may also be hierarchical, nested. This rule may often be optional as just embracing a set of elements in parentheses without any other rule can be treated as a list too, which may also be nested. On default, the remote values by `unit` will be returned as by rule `rake`, but combining it with rules `rake` or `order`, the order of the returned values can be stated explicitly.

unique—returns only unique elements received from the embraced scenario final positions while omitting duplicates (such results may form a list without repeating elements).

sum—returns the sum of all final values of the scenario embraced (modifier `unique` can be used here for summing only unique final values).

count—returns the number of all resultant values associated with the scenario embraced, rather than values themselves as by the previous rules (`unique` can be used too for counting only unique values).

first, **last**, **min**, **max**, **random**, **average**—return, correspondingly, the first, the last, minimum, maximum, randomly chosen, or average value from all terminal values returned by the scenario embraced. The rules first and last may also need ordering of the results similar to rule order discussed before.

element—returns the value of an element of the list on its left operand requested by its index or content (see corresponding usage of rules content and index) given by the right operand. If the right operand is itself a list of indices or contents, the result will be a list of corresponding values from the left operand. If element is used within the left operand of assignment, instead of returning values it will be providing an access to them, in order to be updated, as explained later. Each given index representing unique order can return from the left operand one or none value (the latter if the index exceeds total number of elements), whereas each content in the right operand can return from the left operand none, one, or more elements, as there can be repeating values at the left.

sortup, **sortdown**—return an ordered list of values produced by the embraced scenario operand, starting from minimum or maximum value and ending, correspondingly, with maximum or minimum one.

reverse—changes to the opposite the order of values from the embraced operand.

fromto—returns an ordered list of values by naming its first and last elements as well as step value, thus allowing the next element to be obtained from the previous one. Another modification may take into account the starting element, step value, and the number of needed elements in the list.

add, **subtract**, **multiply**, **divide**, **degree**—perform corresponding operations on two or more operands of the scenario embraced. If the operands themselves represent multiple values, as lists, these operations are performed between peer elements of these lists, with the resulting value being multiple, as a list too.

separate—separates the left operand string value by the string at the right operand used as a delimiter (in hope to be present at the left) in a repeated manner for the left string, with the result being the list of separated values. If the right operand is a list of delimiters, its elements will be used sequentially and cyclically unless the string at the left is fully partitioned. If the left operand represents a list of strings, each one is separated by the right operand as above, with the resultant lists of separated values merged into a common list in the order they were produced.

unite—integrates the list of values at the left (as strings, or to be converted into strings automatically) by a repeated delimiter as a string too (or a cyclic list of them) at the right into a united string.

attach—produces the resultant string by connecting the right string operand to the end of the left one. If operands are lists with more than one element, the attachment is made between their peer elements, receiving the resultant list of united strings. This rule can also operate with more than two operands.

append—forms the resultant list from left and right operands, appending the latter to the end of the former, where both operands may be lists themselves. More than two operands can be used too.

common—returns intersection of two or more lists as operands, with the result including only same elements of all lists, if any, otherwise `nil`.

withdraw—its result will be the first element of the list provided by the embraced operand, which can be a variable too, with withdrawing this element from the list (thus simultaneously changing the content of the variable). This rule can have another operand providing the number of elements to be withdrawn and represented as the result. When the embraced list is empty, the rule returns `nil` value and terminates with `fail` state.

increment—adds 1 to the value of the embraced operand which will be the result on this rule, with simultaneous changing the content of the operand itself (thus having sense only if it is a variable, which will be having now the increased value). If another value, not 1, has to be added, the second operand can be used for keeping this value.

decrement—behaves similar to the previous rule `increment` but subtracts rather than adds 1 from the value of the embraced operand, with the content of the latter simultaneously changed too. Second operand can be used too if the value to be withdrawn not equals 1. In all cases if the result of decrementing appears to be less than zero, the rule will terminate with `fail` and value `nil`.

access—returns a reference to the internal history-based optimized structure (which may be spatially distributed) leading from the rule-activation node to the reached terminal nodes of the embraced scenario. This reference can be remembered in a variable and subsequently used to reach exactly the same terminal nodes again in a most economic and speedy manner. The terminal nodes reentry can be subsequently performed by the rule `follow` described before, with its operand reflecting the remembered access reference got by `access`.

invert—changes the sign of a value or orientation of a link to the opposite, while producing no effect on zero values or non-oriented links.

apply—organizes application of the first operand as one of rules described above or a set of such rules operating jointly (names of which can also be obtained by arbitrary scenario standing for this operand) to the same second scenario operand, which may be arbitrary too. If multiple application rules engaged on the first operand, the obtained results on the second operand can be multiple too.

location—returns world location of the final node reached by the embraced scenario, which will mean for a virtual node its network address, and in case of physical node its physical coordinates. This will be equivalent to using in the final world positions environmental variables ADDRESS or WHERE for providing respective open values, with collecting these values by other echo rules, but directly using `location` may be convenient in certain cases.

3.6.6 Verification

This class of rules has the following main variants.

verification ➜ equal | nonequal | less | lessorequal | more | moreorequal | bigger | smaller | heavier | lighter | longer | shorter | empty | nonempty | belong | notbelong | intersect | notintersect | yes | no

These rules provide control state thru or fail reflecting the result of concrete verification procedure, also nil as own resultant value, while remaining in the same world positions after completion.

equal, nonequal, less, lessorequal, more, moreorequal, bigger, smaller, heavier, lighter, longer, shorter—make corresponding comparison between left and right operands, which can represent information or physical matter/objects, or both. In case of vector operands, state thru appears only if all peer values satisfy the condition set up by the rule (except nonequal, for which even a single non-correspondence between peers will result in overall thru). The list of such rules can be easily extended for more specific applications, if supported properly on the implementation level.

empty, nonempty—checks for emptiness (i.e. non-existence, same as nil) or non-emptiness (existence) of the resultant value obtained from the embraced scenario.

belong, notbelong—verifies whether the left operand value (single or a list, with all elements) belongs as a whole to the right operand, potentially a list too.

intersect, notintersect—verifies whether there are common elements (values) between left and right operands, being generally lists. More than two operands can be used for this rule too, with at least a single same element to be present in all of them to result in thru.

yes—verifies generalized state of the embraced scenario providing own control state thru in case of thru or done from the scenario, and control state fail in case of fail or fatal otherwise.

no—verifies generalized state of the embraced scenario with own control state thru in case of fail or fatal from the scenario, and control state fail in case of thru or done.

3.6.7 Assignment

There are two rules of this class:

assignment ➜ assign | assignpeers

These rules assign the result of the right scenario operand (which may be arbitrarily remote, also represent a list of values) to the variable or set of variables directly named or reached by the left scenario operand, which may be remote too. The left operand can also provide pointers to certain elements of the reached variables which should be changed by the assignment rather than the whole contents of variables (see also rule element mentioned before). These rules will leave control in the same world position they've started, its resultant state thru if assignment was successful otherwise fail, and the same value (which may be a list) as assigned to the left operand. There are two options of the assignment, as follows.

assign—assigns the same value of the right operand (which may be a list) to all values (like, say, node names) or variables accessed (or their particular elements pointed, which may themselves become lists after assignment) by the left operand. If the right operand is represented by nil or empty, the left operand variables as a whole (or only their certain elements pointed) will be removed.

assignpeers—assigns values of different elements of the list on the right operand to different values or variables (or their pointed elements) associated with the destinations reached on the left operand, in a peer-to-peer mode.

3.6.8 Advancement

This class of rules has the following variants:

advancement ➜ advance | slide | repeat | align | fringe

These rules can organize forward, or "in depth" advancement in space and time of the embraced scenarios separated by comma. They can evolve within their sequence in synchronous or asynchronous mode using modifiers synchronous or asynchronous (the second one optional, as asynchronous is default mode).

advance—organizes stepwise scenarios advancement in physical, virtual, executive or combined spaces, also in a pure computational space (the latter when staying in the same world nodes, thus moving in time only). For this, the embraced SGL scenario-operands are used in a sequence (as written) where each new scenario applies from all terminal world points reached by the previous scenario. The final, resultant world positions and values on the whole rule are associated with the final steps of the *last scenario* on the rule (more correctly: of the invocation of all its copies, which may operate in parallel). And the rule's resultant state is a generalization of control states associated with these final steps. If no final steps occur with states thru or done, the whole advancement on this rule is considered as failed (with generalized state fail), thus resulting without possibility to continue scenario evolution in this direction.

On default or with modifier asynchronous, the sequence of scenarios on advance develops in space and time independently in different directions, with a new scenario starting immediately in the points reached by the previous scenario. This means that different, not the same, operand scenarios in their sequence may happen to be active simultaneously at the same time, as having developed independently and in parallel, with different times of their completion. With the use of synchronous modifier, all invocations of every new scenario (in general: all its multiple copies) in their sequence can start only *after full completion* of all invocations of the previous scenario.

slide—works similar to the previous rule unless a scenario fails to produce resultant state `thru` or `done` from some world node. In this case the next scenario from their sequence will be applied from the same starting position of the previous, failed scenario and so on. The resultant world nodes and values in them will be from the last successfully applied scenarios (not necessarily the same in their sequence, as independently developing in different directions). The results on the whole rule, in their extreme, may even happen to correspond to the existing value of the node in which the rule started (including this node's world position), with state `thru` always being the resultant state in any cases. Both synchronous and asynchronous modes of parallel interpretation of this rule, similar to the previous rule `advance`, are possible, where in the synchronous option, different scenarios (not necessarily their same copies) can simultaneously start only after full completion of the previous parallel steps (also potentially involving different scenarios).

repeat—invokes the embraced scenario as many times as possible, with each new iteration taking place from all final positions with state `thru` reached by the previous invocation. If a scenario iteration fails, its starting position with its value will be included into the set of final positions and values on the whole rule (this set may have starting positions from different failed iterations which developed independently in a distributed space). Similar to the previous rule `slide`, in the extreme case, the final set of positions on the whole rule may happen to contain only the position from which the rule started, with state `thru` and value it had at the beginning. By supplying additional numeric modifier to this rule, it is possible to explicitly limit the number of allowed scenario repetitions. Of course, the operand-scenario can be easily internally organized to properly control the allowed number of iterations itself, but using this additional modifier may be useful in some cases.

Both synchronous and asynchronous modes of parallel interpretation of this rule similar to the previous rules `advance` and `slide` are possible. In the synchronous mode, at any moment of time only the same scenario iterations can develop (whereas some previous ones may have already stopped in other directions). In the asynchronous case, there may be different iterations working in parallel.

align—is based on confirmation of full termination of all activities of the embraced operand-scenario in all its final nodes. Only after this, the remaining scenario part, if any, will be allowed to continue from all the nodes reached.

fringe—allows us to establish certain constraints on the terminal world nodes reached by the embraced scenario with final values in them, to be considered as starting positions for the following scenario parts. For example, by comparing values in all terminal nodes and allowing the scenario to continue from a node with maximum or minimum value, integrating this rule with previously mentioned rules, like: **max_fringe** or **min_fringe**.

For the advancement rules, frontal variables propagate on the forefronts together with advancement of control and operations in distributed spaces, with next scenarios or their iterations picking up frontal variables brought to their starting points by the previous scenarios or their iterations, being also replicated if this control splits into different branches. And the capability and variants of splitting into branches will be considered in detail in the following section.

3.6.9 Branching

There rules from this class are as follows:

branching ➜ branch | sequence | parallel | if | or | and | choose |
quickest | cycle | loop | sling | whirl | split

These rules allow the embraced set of scenario operands to develop "in breadth", each from the same starting position, with the resultant set of positions and order of their appearance depending on the logic of a concrete branching rule.

Branching may be static and explicit if we have a clear set of individual operand scenarios separated by comma. It can also be implicit and dynamic, as explained later. For all branching rules that follow, the frontal variables associated with the rule's starting position will be replicated together with their contents and used independently within different branches. Details of this replication for frontal variables with physical matter rather than information can depend on application and implementation details.

An explanation of how these rules work is as follows.

branch—the most general and neutral variant of branching, with logical independence of scenario operands from each other and any possible order of their invocation and development from the starting position (say, ranging from chaotic to strictly sequential to fully parallel, also to any mixture thereof). The resultant set of positions reached with their associated values will unite all terminal positions & values on all scenario operands involved. The resultant control state on the whole rule will be based on generalization of generalized control states on all scenario branches (based on powers of control states, i.e. from max to min as fatal, thru, done and fail, as mentioned before).

sequence—organizing strictly sequential invocation of all scenario operands, regardless of their resultant generalized control states, and launching the next scenario only after full completion of the previous one, regardless of its success or failure. The resultant set of positions, values, and rule's control state will be similar to branch.

parallel—organizing fully parallel development of all scenario operands from the same starting position (at least as much as this can be achieved within the existing environment, resources, and implementation). The resultant set of positions, values, and rule's control state will be similar to the previous two rules.

if—usually has three scenario operands. If the *first* scenario results with generalized termination state thru or done, the *second* scenario is activated, otherwise the *third* one will be launched. The resultant set of positions & associated values will be the same as achieved by the second or third scenarios after their completion. If the third operand-scenario is absent and the first one results with fail, or only the first operand is present regardless of its success or failure, the resultant position will be the one the rule started from, with state thru and value it had at the start.

or—allows *only one* operand scenario with the resulting state thru or done, without any predetermined order of their invocation, to be registered as resultant, with the final positions & associated values on it to be the resulting ones on the whole rule. The activities of all other scenario operands and all results produced by them will be terminated and cancelled. If no branch results with thru or done, the rule will terminate with fail and nil value. If used in combination with the previous rules sequence and parallel, it may have the following peculiarities.

    or_sequence—will launch the scenario operands in a strictly sequential manner, one after the other as they are written, waiting for their full completion before activating the next operand, unless the first one replies with generalized state thru or done (providing the result on the rule as a whole). Invocation of the remaining scenarios in the sequence will be skipped.

    or_parallel—activates all scenario operands in parallel from the same current position, with the first one in time replying with generalized thru or done being registered as the resultant branch for the rule. All other branches will be forcefully terminated without waiting for their completion (or just ignored, depending on implementation, which in general may not be the same due to side effects when working with common resources).

The resultant scenario chosen in all three cases above provides its final set of positions with values and states in them as the result on the whole rule. If no scenario operand returns states thru or done, the whole rule will result with state fail in its starting position and nil as resultant value.

    and—activates all scenario operands from the same position, without any predetermined order, demanding all of them to return generalized states thru or done. If at least a single operand returns generalized fail, the whole rule results with state fail and nil value in the starting position while terminating the development of all other branches, which may still be in progress. If all operand scenarios succeed, the resulting set of positions unites all resultant positions on all scenario-operands with their associated values. Combining rule and with rules sequence and parallel (as we did for or) will clarify their activation and termination order, as follows. (These two options can, in principle, produce differing general results if different scenario operands work with intersecting domains and share intermediate results.)

    and_sequence—activates scenario-operands from the same position in the written order, launching next scenario only after the previous one completes, and terminating the whole rule when the current scenario results with fail. The remaining operands will be ignored, and all results produced by this and all previous operands will be removed (as far as this can be achievable in a distributed environment).

    and_parallel—activates in parallel all scenario operands from the same world position, terminating the rule when the first one in time results with fail, while aborting activity of all other operands and removing all results produced by the rule.

    choose—chooses a scenario branch in their sequence *before* its execution, using certain additional parameters among which, for example, may be its numerical order in the sequence (or a list of such orders to select more than one branch). This rule can also be aggregated with other rules like first, last or random, i.e. forming combined rules: **choose_first**, **choose_last**, **choose_random**. The resultant set of positions on the rule, their values and states will be taken from the branch (or branches) chosen.

quickest—selects the first branch in time replying its complete termination, regardless of its generalized termination state, which may happen to be `fail` too, even though other branches (to be forcefully terminated now) could respond later with `thru` or `done`. The state, set of positions on this selected branch, and their associated values (if any) will be taken as those for the whole rule. (This rule assumes that different branches are launched independently and in parallel.)

It differs fundamentally from the rule `or_parallel` as the latter selects the first in time branch replying with success (i.e. `thru` or `done`) for which, in the worst case, all branches may need to be executed in full to find the branch needed. A modification of `quickest` may have an additional parameter establishing, for example, time limit within which replies are expected or allowed from branches (and there may be more than one branch as the result). Otherwise the rule will terminate with failure if no branch responds in time.

cycle—repeatedly invokes the embraced scenario from the same starting position until its resultant generalized state remains `thru` or `done`, where on different invocations same or different sets of resultant positions (with same or different values) may emerge. The resultant set of positions on the rule will be an integration of all positions on all successful scenario invocations with their associated values. The following scenario will be developing from all these world positions reached (some or all may be repeating) except the ones resulting with state `done`. If no invocation of the embraced scenario succeeds, the resultant state `fail` in the starting position with `nil` value will emerge.

loop—differs from the previous rule in that the resultant set of positions on it being only the set produced by the *last* successful invocation of the embraced scenario (the rule will terminate, as before, with `fail` and `nil` in the starting position if no invocation succeeds).

sling—invokes repeatedly the embraced scenario until it provides state `thru` or `done`, always resulting in the same starting position with state `thru` and its previously associated value when the last iteration results with `fail`.

whirl—endlessly repeating the embraced scenario from the starting position regardless of its success or failure and ignoring any resultant positions or values produced. External forceful termination of this construct may be needed, like using first in time termination of a competitive branch (say, under the higher-level rule `or_parallel`).

It could also be possible to set an explicit limit on the number of possible repetitions (or duration time) in the above mentioned cycling-looping-slinging-whirling rules—by supplying them with an additional parameter restricting the repeated scenario invocations.

split—performs, if needed, additional static or dynamic partitioning of the embraced scenario to different branches, especially in complex and not clear at first sight cases, all starting from the same current position. It may be used alone or in combination with the above mentioned branching rules, preparing separate branches for these rules, ahead of their invocation. Some examples follow.

- If `split` embraces explicit branches separated by commas, it does nothing as the branches are already declared.
- It the embraced single operand represents broadcasting move or hop (creative or destructive including) in multiple directions, the branches are formed from all possible variants of elementary moves or hops, *before* their execution.
- If the rule's operand is an arbitrary scenario (not belonging to the two cases above), the branches are formed *after* its completion, where each position reached by the scenario (with associated values) starts a new branch.
- If arbitrary scenario terminates with a single position which has associated multiple values (i.e. a list), each constituent value in such a list starts an individual branch, becoming a starting value of this branch.

3.6.10 Transference

There are two rules of this class:

transference ➜ run | call

They organize transference of control in distributed scenarios.

run—transfers control to the SGL code treated as a procedure and being a result of invocation of the embraced scenario (which can be of arbitrary complexity and space coverage, or can just be an explicit constant or variable). The procedure (or a list of them) obtained and activated in such a way can produce a set of world positions with associated values and control states as the result on the rule, similar to other rules. In case of failure to treat and activate results of the embraced operand as an SGL scenario, this rule will terminate with value `nil` and state `fail` in the node it started.

call—transfers control to the code produced by the embraced scenario which may represent activation of external systems (including those working in other formalisms). The resultant world position on `call` will be the same where the rule started, with value in it corresponding to what has been returned from the external call and state `thru` if the call was successful, otherwise `nil` and `fail` of the latter two.

3.6.11 Exchange

exchange ➜ input | output | send | receive | emit | get

input—provides input of external information or physical matter (objects) on the initiative of SGL scenario, resulting in the same position but with value received from the outside. The rule may have an additional argument clarifying a particular external source from which the input should take place. The rule extends possibilities provided by reading from environmental variable IN explained before.

output—outputs the resultant value obtained by the embraced scenario, which can be multiple, with the same resultant position as before and associated value sent outside (in case of physical matter, the resultant value may depend on the applications). The rule may have an additional pointer to a particular external sink. The rule extends possibilities provided by assignment to the previously explained environmental variable OUT.

send—staying in the current position associated with physical, virtual, executive (or combined) node, sends information or matter obtained by the scenario on the first operand to other similar node given by name, address or coordinates provided by the second operand, assuming that a companion rule receive is engaged there. The rule may have an additional parameter setting acceptable time delay for the consumption of this data at the receiving end. If the transaction is successful, the resultant position will be the same where the rule started with state thru and value sent (in case of physical matter, this may depend on application and implementation capabilities), otherwise nil and state fail.

receive—a companion to rule send, naming the source of data to be received from (defined similarly to the destination node in send). Additional timing (as a second operand) may be set too, after expiration of which the rule will be considered as failed. In case of successful receipt of the data, the rule will result in the same world position and the value obtained (information or matter) from send and state thru, otherwise will terminate with value nil and state fail.

emit—depending on implementation and technical capabilities, can trigger non-local or global continuous broadcasting of the data obtained by the embraced scenario, possibly, with tagging of this source (like setting the emission frequency). Another operand providing time allowed for this broadcasting may be present too. No feedback from possible consumers of the sent data is expected. Will terminate in the application node with the broadcast value and state thru in case of success, otherwise with nil and fail.

get—tries to receive data which can be broadcast from some source (say, identified by its tag or frequency), with resultant value as the received data and state thru in the application node, otherwise with nil and state fail. Similar to the previous rule, additional operand can be introduced for limiting the activity time of this rule. No synchronization with the data emitting node is expected.

3.6.12   Timing

The following two options are available for this rule:

timing ➜ sleep | allowed

These rules are dealing with conditions related to time interval for the scenarios they embrace.

sleep—establishes time delay defined by the embraced scenario operand, with suspending activities of this particular scenario branch in the current node. The rule's starting position and its existing value, also state thru, will be the result on the rule after the time passed. Similar time delay of the related branch can also be achieved by assigning the current absolute time (say, received from environmental variable TIME) incremented by the needed delay value to environmental variable WHEN described before.

allowed—sets time limit by the first operand for an activity of the scenario on the second operand. If the scenario terminates before expiration of this time frame, its resultant positions with values and states will define the result on this rule. Otherwise the scenario will be forcibly aborted, with state fail and value nil as the rule's result in its starting position.

3.6.13   Qualification

This class containing the following rules:

qualification ➜ contain | release | free | blind | quit | abort | stay | lift | seize

These rules are providing certain qualities or abilities, also setting constraints or restrictions to the scenarios they embrace, as follows.

contain—restricts the spread of abortive consequences caused by control state fatal within the ruled scenario. The latter state may appear automatically and accidentally in different points or can be assigned explicitly to environmental variable STATE, triggering emergent completion of all scenario processes and removal of temporary data associated with them. The resultant position will be the one the rule started from, its value nil, and state fail. Without occurrence of fatal, the resultant positions, their values and states on the rule will be exactly the same as from the scenario embraced.

The destructive influence from state fatal is also automatically stopped if the scenario in which it may appear is covered by rule state (converting any embraced control state into a value), also rules yes and no (first changing the embraced state into fail and second into thru in case of fatal), as described before. But after these three rules the resultant world positions always correspond to the single rule's starting node regardless of what scenario produces, whereas rule contain results in exactly the same final nodes (which may be many) as the scenario embraced.

release—allows the embraced scenario develop free from the main scenario, abandoning bilateral control links with it starting from the application position (the main scenario after the rule's activation "will not see" this construct any more). The released, now independent scenario will develop using standard subordination and command & control mechanisms as usual. For the main scenario, this rule will immediately result in its starting position with state thru and original value there.

free—differs from the previous case in that despite its independence and control freedom from the main scenario, as before, the embraced scenario will nevertheless be obliged to return the final data obtained in its terminal positions to the main scenario if such a request issued by rules of the latter.

blind, quit, abort—after full completion of the embraced scenario, these rules result in the same position the rule started and respectively states done, fail, or fatal, thus preventing further scenario development from this point (also causing nonlocal processes and results termination in case of fatal). These rules represent more economic solutions than planned termination of all final branches of the embraced scenario with these sates. If the ruled scenario is omitted (i.e. the rules names standing alone), these rules will be equivalent to assigning the mentioned above states to environmental variable STATE in the position they started.

stay—whatever the scenario embraced and its evolution in space, the resultant position will always be the same this rule started from (and not the termination positions of the ruled scenario), with value nil and state thru in it. If the ruled scenario is omitted, this rule standing alone just represents an empty operation in the current point or assignment state thru to variable STATE in it. Such empty operation can also be used for declaring new scenario branch which may be used by the rest of the scenario starting from this point.

lift—lifts blocking of further scenario developments set up by states done in the embraced scenario wherever it happened (including equivalent effect caused by rules blind), substituting them with thru and allowing further developments from all such positions, which may massive and space-distributed.

seize—establishes, or "seizes", an absolute control over the resources associated with the current virtual, physical, executive or combined node, blocking these from any other accesses and allowing only the embraced scenario to work with them (thus preventing possible competition for the node's assets which may lead to unexpected results). This resource blockage is automatically lifted after the embraced scenario terminates. The resultant set of positions on the rule with their values and states will be the ones from the scenario embraced. If the node has already been blocked by other scenario exercising its own rule seize, the current scenario will be waiting for the release of the node. If more than two scenarios are competing for the node's resources, they will be organized in a FIFO manner at the node.

3.6.14 Grasp

Rule's identifier can be expressed not only by a directly given name but also by the name-interpreted result produced by a scenario of any complexity. It can also be a compound one, integrated from multiple scenario-produced values, so in general, we may have the following:

rule ➜ *grasp* ➜ *constant* | *variable* | *rule* ({*grasp*, })

Under this extended definition, resulting from recursive SGL syntax, additional parameters can be associated with the rule names, before embracing the main scenario operands. Such an aggregation can simplify the structures of SGL scenarios, also making them flexible and adjustable to changing goals and environments in which they operate.

3.7 Possible Scenario Simplifications

Certain simplifications and flexibilities for writing scenarios may be allowed, which can be handled by extended SGL interpreter implementations, or initial, before execution, preprocessing of such scenarios with conversion of "alien" constructs into the native SGL syntax, or both. These cases may also relate to using conventional representation of operations and delimiters, like in other languages, while generally remaining within the recursive syntax described in this chapter. Let us consider some simple examples.

Example 1 Instead of using the stepwise propagation rule advance, we may just separate its scenario operands by traditional semicolon with or without the embracing opening-closing parentheses, which may be needed for the overall scenario structuring.

```
advance(scenario1, scenario2, scenario3) =>
(scenario1; scenario2; scenario3) =>
scenario1; scenario2; scenario3
```

Example 2 Instead of using the most general branching rule branch, allowing us to independently evolve from the current world point in different directions, we may just omit it, leaving the scenario operands separated by commas (with or without the embracing parentheses, depending on the overall scenario structuring).

```
branch(scenario1, scenario2, scenario3) =>
(scenario1, scenario2, scenario3) =>
scenario1, scenario2, scenario3
```

Example 3 In case of combination of rules advance and branch with the use of simplified representations with their omission, we may leave or omit the embracing parentheses, taking into account that comma as a delimiter (also as operation in our case) is superior to semicolon.

```
advance(branch(scenario1, scenario2), scenario3) =>
((scenario1, scenario2); scenario3) =>
(scenario1, scenario2); scenario3 =>
scenario1, scenario2; scenario3
```

Example 4 For rule `create` forming new link-node pair if link name is to be its first operand and node name second, as usual, these names may be used without rules explicitly identifying them, i.e. `link` and `node`. Also, if link and/or node names do not intersect with special SGL constants or representation of variables, single quotes embracing them may be omitted (with caution, however, for not making mistakes).

```
create(link('Linkname'), node('Nodename')) =>
create('Linkname', 'Nodename') =>
create(Linkname, Nodename)
```

Example 5 Similar simplifications can also be made for rule `hop` setting propagation via a link with proper name to a named node. Also, instead of word `hop`, we may use symbol # with link and node manes standing on its different sides.

```
hop(link('Linkname'), node('Nodename')) =>
hop('Linkname', 'Nodename') =>
hop(Linkname, Nodename) =>
Linkname # Nodename
```

Example 6 In case of direct hop to a node by its name only, expressed by #, the special parameter `direct`, standing instead of link name, may be omitted (as the left operand of hop).

```
direct #'Nodename' => #'Nodename' => # Nodename
```

Example 7 In case of direct hop to any node by link name only, expressed by #, the special parameter `any`, standing instead of the node name, may be omitted (as the right operand of hop).

```
'Linkname'# any => 'Linkname'# => Linkname #
```

Example 8 Traditional symbols for different numerical, string, comparison, etc. operations can be used instead of rules. Some variants follow.

```
add(25.0, 78.66, 9) => 25.0 + 78.66 + 9
greater(105, 36) => 105 > 36
attach('Peter',' and ','Paul') =>
```

   `'Peter'&'and '&'Paul'`, resulting in: `'Peter and Paul'`

   `append('Peter', 'Paul')=>'Peter' && 'Paul'`, resulting in: (`'Peter'`, `'Paul'`), or if parentheses not needed for structuring and quotes being redundant, say, as for an output: `Peter, Paul`.

Example 9 Rule `unit`, as already mentioned, can be omitted when grouping elements (say, constants) into lists, which may be nested, with just parentheses being sufficient.

```
unit(constant1, constant2,
     unit(constant3, constant4)) =>
(constant1, constant2, (constant3, constant4))
```

3.8 Conclusion

We have provided full details of the Spatial Grasp Language, SGL, suitable for parallel processing in large distributed environment both virtual - networked and physical, which can be arbitrarily large and have no borders. Having only three conceptual components like constants, variables and rules, and universal recursive syntax it allows us to describe and organize arbitrary complex processes in a variety of distributed systems. The presented latest and updated version is particularly suitable for dealing with social systems and their expression by large distributed social networks discussed in this book.

The language allows us to grasp complex problems in different spaces and their solutions on topmost, semantic level in pattern-matching mode, while allowing at the same time to describe and deal with any details needed, on all levels, which can make SGL basic if not the single language for advanced systems dealing distributed graphs and networks without any centralized resources. The peculiarity of high-level language constructs and their combinations allows us to shift most of system management routines to the automatic and intelligent interpretation level. The language space-grasping philosophy and organization also allows it to be easily extendable to any other classes of problems by simply adding new specific rules within the same recursive syntax.

References

1. P. Sapaty, *Mobile Processing in Distributed and Open Environments* (Wiley, New York, 1999)
2. P. Sapaty, *Ruling Distributed Dynamic Worlds* (Wiley, New York, 2005)
3. P. Sapaty, Managing Distributed Dynamic Systems with Spatial Grasp Technology (Springer, Heidelberg, 2017)
4. P. Sapaty, A brief introduction to the spatial grasp language (SGL). J. Comput. Sci. Syst. Biol. **9**(2) (2016)
5. P.S. Sapaty, Spatial grasp language for distributed management and control. Math. Mach. Syst. (3) (2016)
6. P. Sapaty, Spatial grasp language (SGL). Adv. Image Video Process. **4**(1) (2016)
7. P.S. Sapaty, Spatial grasp language (SGL) for distributed management and control. J. Robot. Netw. Artif. Life **4**(2) (2016)
8. P.S. Sapaty, M. Sugisaka, A language for programming distributed multi-robot systems, in Proceedings of the Seventh International Symposium on Artificial Life and Robotics (AROB 7th '02), B-Com Plaza, Beppu, Oita, Japan, pp. 586–589, 16–18 January 2002
9. P.S. Sapaty, Mobile programming in WAVE. Math. Mach. Syst. **1**, 3–31 (1998)
10. P.S. Sapaty, P.M. Borst, An overview of the WAVE language and system for distributed processing in open networks, Technical Report, Dept. Electronic & Electrical Eng, University of Surrey, June 1994
11. P.S. Sapaty, A brief introduction to the WAVE language, Report No. 3/93, Faculty of Informatics, University of Karlsruhe (1993)
12. P.S. Sapaty, WAVE-1: a new ideology of parallel processing on graphs and networks. Future Gener. Comput. Syst. **4** (North-Holland) 1988)
13. P.S. Sapaty, The WAVE-1: a new ideology and language of distributed processing on graphs and networks. Comput. Artif. Intell. (5) (1987)
14. P.S. Sapaty, A wave language for parallel processing of semantic networks. Comput. Artif. Intell. **5**(4) (1986)
15. P.S. Sapaty, The WAVE-0 language as a framework of navigational structures for knowledge bases using semantic networks, in *Proceedings USSR Academy of Sciences*. Technical Cybernetics, No. 5 (1986) (in Russian)
16. P.S. Sapaty, A wave approach to the languages for semantic networks processing, in *Proceedings International Workshop on Knowledge Representation. Section 1: Artificial Intelligence* (Kiev, 1984) (in Russian)
17. P.S. Sapaty, On possibilities of the organization of a direct intercomputer dialogue in ANALYTIC and FORTRAN languages, Publ. No. 74-29 (Inst. of Cybernetics Press, Kiev, 1974) (in Russian)

Chapter 4
Distributed Network Processing Basics

4.1 Introduction

Examples of some basic network analysis and processing mechanisms expressed in SGL and oriented on parallel and fully distributed operation on networks of arbitrary size, structure and space coverage will be presented and explained. These include initial settling in network nodes, parallel and stepwise movements through network structure, calculation of distributed network elements, forming breadth-first and depth-first spanning trees and finding paths through network nodes and links, shortest ones including, also initial creation and modification of arbitrary networks in distributed spaces.

These mechanisms and solutions presented in different language permitted modifications will be used in numerous SGL scenarios throughout the rest of this book as basic ones for dealing with distributed networks in a self-spreading and pattern-matching mode, and should not be, if possible, explained again.

The solutions shown in this chapter are based on previous applications of this distributed pattern-matching approach for solving graph and network problems which are summarized in the previous books [1–3], with some being initially influenced by echo algorithms on graphs [4].

4.2 Notes from Graph Theory

Many fundamental concepts and metrics in social network analysis are derived from graph theory [5–7] which is formally representing social networks with their structural properties [8–14]. Some of these concepts are mentioned below, to be subsequently used in other chapters.

Node degree. In graph theory, the degree of a node in a graph is the number of edges incident to the node. There are at most $N*(N-1)/2$ edges for an undirected

© Springer Nature Switzerland AG 2019

P. S. Sapaty, *Holistic Analysis and Management of Distributed Social Systems*, Studies in Systems, Decision and Control 184, https://doi.org/10.1007/978-3-030-01830-6_4

graph and at most N*(N − 1) edges for a directed graph, where N is the number of nodes. (These estimates assume that any two nodes of the graph are connected by no more than a single edge.)

Node density. The density of an undirected graph can be defined as (2*E)/N*(N − 1), where E is the number of edges. On the other hand, the density of a directed graph can be defined as E/N*(N − 1).

Path length. The path length is the number of edges in the sequence that a walk follows. In a path, all nodes and edges appear only once in the sequence. Therefore, the path length can be defined as the distances between pairs of nodes in a network graph, and average path length is the average of these distances between all pairs of nodes.

Component size. When the component size is concerned, a connected graph needs to be discovered first since the component size is counted by the number of connected nodes in a graph. A graph is connected if all pairs of nodes are reachable, and for each pair of two nodes, one of them is reachable from the other. On the other hand, if a graph is not connected, the graph can be partitioned into several connected subgraphs where each component size can be calculated by the number of connected nodes in each subgraph.

Geodesics. Network metrics are often measured using *geodesics*—or shortest paths. They make the (often erroneous) assumption that all information/influence flows along the network's shortest paths only. But the networks often operate via both direct and indirect paths.

4.3   Exemplary Network Representation

We will be using here, just for convenience, an exemplary network topology known as "kite graph" [8, 9] and shown in Fig. 4.1 (with our own names of nodes and arcs), which is often exhibited when analysing social networks [10–13].

The following sections are showing and explaining different scenarios on this network topology, expressed in SGL, which are operating directly on the network surface in a smog-, flood-, or virus-like parallel and distributed mode. The self-evolving scenarios are covering and matching the network topology which can be arbitrarily distributed in space without any boundaries.

4.4   Landing on a Network and Elementary Moves on It

This set of initial network operations will be including the following ones.

- Hopping directly to some node with a given *nodename* from outside the network, and reducing the scenario text wherever possible while preserving its unambiguity:

```
hop(direct, node('nodename'))
```

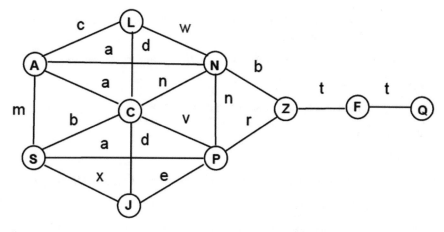

Fig. 4.1 Exemplary network structure

or, after integrating the rule's name with modifiers:

```
hop_direct_node('nodename')
```

next, as the node name is clearly used here and nothing else:

```
hop_direct('nodename')
```

or, as only node can be mentioned in a direct hop:

```
hop_node('nodename')
```

next, if *nodename* does not intersect with other SGL constructs:

```
hop_node(nodename)
```

next, with further simplification, as only node name can be used in a direct hop:

```
hop(nodename)
```

and finally, as the shortest possible expression:

```
#nodename
```

After hoping to the destination node the scenario control will be residing there.

- Hopping directly to some node A from any other node of the network while staying in it, will be similar to the hop to A from the network's outside, as before, with scenario control moving into A too:

```
hop_direct_node('A')
```

up to the shortest variant:

```
#A
```

- Hopping directly to node A and then from it directly to node B, with scenario control finally moving into B too:

      advance(hop_direct_node('A'), hop_direct_node('B'))

  or:

      hop_direct_node(A); hop_direct_node(B)

  up to the shortest version:

      #A;#B

- Hopping from the current node by the named link to the named neighboring node:

      hop(link('linkname'), node('nodename'))

  The shortest version when link and node names are used in their standard sequence (first link then node), also both link and node names are not interfering with other SGL constructs, will be as follows:

      linkname#nodename

  If link to be used is named a, and target node name is R, this will be:

      a#R

- Hopping by the named link(s) to any neighbouring nodes these links may lead to, when staying in some node:

      hop(link(linkname), node(any))

  or, by just omitting information about destination nodes at all, as these can have any names:

      hop_links(linkname)

  or as shortest:

      linkname#

  Instead of modifier any another modifier all can be used too, as links with the given name will be tried to be traversed to all possible neighbouring nodes.
  As for Fig. 4.1, when staying in node A and hopping from it by all links named a, as follows:

      a#

  will lead us to nodes C and N with subsequent staying in both of them.

- Hopping by any link to a particular neighbouring node

      hop(links(any), node(nodename))

  For example, the staying in node A in Fig. 4.1 and hopping by any existing link to node S can be expressed by the following:

```
hop(links(any), node(S))
```

or just by:

```
any#S
```

which will pass the link m before arriving in node S.

- Hopping by all links to a particular neighbouring node:

```
hop(link(all), node(nodename))
```

The result could differ from the previous one when hopping to node S if more than a single link led to this node from the current one—this node would then be entered more than once, with multiple resultant control branches associated with the same destination node.

- Hopping to all neighbouring nodes via all links leading to them, regardless of link names used:

```
hop(links(all), node(any))
```

or:

```
hop_links(all)
```

or as the shortest:

```
all#
```

Using this operation when staying in node A will lead to nodes L, N, C, S.

- To get a result similar to the previous one, we may tell in A that all neighbouring nodes should be directly reached:

```
hop_direct(neighbors)
```

or as shortest:

```
#neighbors
```

The result may, however, differ from the previous case, where neighbouring nodes can be reached more than once if multiple links lead to them, whereas in the current case each neighbouring node will be entered only once. This direct hop may also be more expensive than reaching the neighbors through links leading to them.

- Direct hopping to all other nodes in the network when staying in some node:

```
hop_direct_nodes(all_other)
```

or:

```
hop_nodes(all_other)
```

or:

```
hop(all_other)
```

or as shortest:

```
#all_other
```

If to apply this when staying in node A of Fig. 4.1, all other network nodes will be entered regardless of existence or non existence of links to and between them, with the resultant scenario control moving in parallel to all these nodes:

```
L, C, N, S, J, P, Z, F, Q
```

- Direct hopping to all nodes in the network including yourself:

```
hop_direct_nodes(all)
```

or:

```
hop_nodes(all)
```

or just:

```
#all
```

Applying this from any node, when staying in it, all network nodes will be directly reached, itself including, regardless of the existence of network links. Same result will be achieved if this operation is applied from the network's outside. For example, all nodes of Fig. 4.1 will be entered in both cases, with the resultant control staying in all of them in parallel.

Direct re-entry of the current node, when staying in it, can be expressed as follows:

```
hop_direct_node(current)
```

or:

```
hop_node(current)
```

or:

```
hop(current)
```

or as shortest:

```
#current
```

4.5   Counting Network Elements

- Counting the number of adjacent links from the current node, with the following options possible:

For links with any orientation or non-oriented links:

```
count(hop_links(all))
```

For oriented incoming links:

```
count(hop_links(-all))
```

For oriented outgoing links:

```
count(hop_links(+all))
```

The shortened versions for the above three cases will be as follows:

```
count(all#), count(-all#), count(+all#)
```

For example, for Fig. 4.1 with all non-oriented links by:

```
count(all#)
```

will receive 6 when staying in node C, and when staying in nodes P or N will get 5.

- Counting the number of all nodes in the network, itself including, when staying in some node or from the network's outside:

```
count(hop_direct_nodes(all))
```

or:

```
count(#all)
```

Result for the network of Fig. 4.1 will be 10.

- Counting all other nodes, when staying in any node:

```
count(hop_direct_nodes(allother))
```

or

```
count(#allother)
```

This will produce result 9 for the network of Fig. 4.1.

- Counting the number of all network links when staying in any node or from the network's outside.

For outgoing oriented links:

```
count(hop_nodes(all); hop_links(+all))
```

For incoming oriented links:

```
count(hop_nodes(all); hop_links(-all))
```

For non-oriented links or links with any orientation:

```
count(hop_nodes(all); hop_links(all))/2
```

or, by comparing node addresses to remove duplicates:

```
count(hop_nodes(all); hop_links(all);
```

```
                    PREVIOUS  >  ADDRESS)
```

or, if node names are unique for removing duplicates too:

```
count(hop_direct(all); hop_links(all);
       PREDECESSOR  >  NAME)
```

We could also use less (<) instead of more (>) for the previous two examples. Examples of shortest versions for all cases mentioned above:

```
count(#all; + all#)
count(#all;-all#)
count(#all;all#)/2
count(#all;all#;PREV > ADDR)
count(#all;all#;PRED > NAME)
```

Only results from the latest three cases for the network of Fig. 4.1 will be valid and equal 18, because mentioning orientation for non-oriented links, with the latter being all in the figure, causes no effect, and the links will be passed in both directions when traversed from adjacent nodes. So the first two cases will result with double number of links, i.e. 36.

- Finding network node with maximum number of links, itself including, with the following options possible.

Obtaining resultant node name together with the number of its links in the reverse order:

```
max(hop_nodes(all);
     append(count(hop_links(all)), NAME))
```

With the shortest version:

```
max(#all;count(all#)&&NAME)
```

The result for the network of Fig. 4.1, if to stay in any node or outside, will be as follows:

```
(6, C)
```

where spatial rule max in SGL compares not only single values but also lists of values coming from all nodes (consisting of two values in our case, with the first one as the number of links of a node, and second as node's name), and the comparison is made by the first item of the lists. If to get only name of the node having maximum links, we may write:

```
element(max(hop_nodes(all);
        count(hop_links(all)) && NAME), 2)
```

With shortened version we will have (where the colon is used instead of rule element):

```
max(#all;count(all#)&&NAME):2
```

The result will just be:

```
C
```

4.6 Compound Moves on a Network

- Organizing succeeding hops in a network from a node to some other node, and then to another one, by giving link and node names for such moves:

```
advance(hop(link(linkname1), node(nodename1)),
        hop(link(linkname2), node(nodename2)))
```

or:

```
hop(link(linkname1), node(nodename1));
hop(link(linkname2), node(nodename2))
```

or, as shortest vesion:

```
linkname1#nodename1;linkname2#nodename2
```

If we are staying in node A of Fig. 4.1 and want to move via link a to node C, and then via link v to node P, we may write:

```
hop(link(a), node(C)); hop(link(v), node(P))
```

or, as shortest:

```
a#C;v#P
```

After reaching node C, the scenario will lose it first part a#c as not needed any more and will continue further with the rest as: v#p, with the resultant control in node P and empty scenario text.

- Repeating succeeding hops from node A along links a, m, n, r, t as far as possible while blocking possible return to the already visited nodes, using for this nodal variable Mark in every node reached.

```
nodal(Mark);
repeat(Mark == nil; Mark = 1;
       hop(links(a, m, n, r, t), nodes(any)))
```

Or, by using special internal interpretation mechanisms implicitly preventing looping and which are triggered in each node when moving to it with rule hopfirst, this may be simplified (the starting node should be re-entered by it too):

```
hopfirst(current);
repeat(hopfirst(links(a, m, n, r, t), nodes(any)))
```

Nodes (N, S, P, C, Z, F) will be visited each time.

The scenario body part covered with rule repeat each time propagates between the nodes reached unless further movement to new nodes by the links named becomes impossible. Also, if from a node more than one link emerges to yet non-visited nodes, the repeat-covered body replicates and moves in networking space to all destination nodes independently.

- Repeating succeeding hops as in the previous case but limited with allowed number of repetitions, say, remembered in frontal variable `Limit`:

```
frontal(Limit) = 2; hopfirst(current);
repeat(decrement(Limit) >= 0;
          hopfirst(links(a, m, n, r, t), node(any)))
```

By directly integrating rule `repeat` with an additional parameter in frontal variable `Limit`, we can explicitly restrict the allowed number of such successive repetitive hops between nodes:

```
frontal(Limit) = 2; hopfirst(current);
repeat_Limit(
 hopfirst(links(a, m, n, r, t), nodes(any)))
```

This can also be achieved by extending the rule's name just with direct constant:
`hopfirst(current);`

```
repeat_2(hopfirst(links(a, m, n, r, t), nodes(any)))
```

Only `nodes(N, S, P, C)` will be visited for the latest three cases.

4.7 Spanning Trees Coverage

Different strategies of covering all nodes of the network when starting from some node will be discussed.

- Reaching all nodes by self-navigating asynchronous spanning tree-coverage scenario.

In the following scenario, starting from certain node and spreading in a combined breadth-depth asynchronous mode, any cycles are explicitly blocked by marking already passed nodes with the help of a nodal variable (let it be `Mark` again) existing in each node.

```
nodal(Mark); Mark = 1;
repeat(hop(links(all), nodes(any));
      empty(Mark); Mark = 1)
```

Or with implicit blocking of cycles by internal node markings:

```
hopfirst(current);
repeat(hopfirst(links(all), nodes(any)))
```

As nodes may have any names, we can just omit explicit mentioning them at all:

```
hopfirst(current);
repeat(hopfirst_links(all))
```

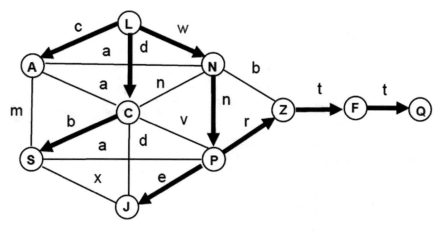

Fig. 4.2 Possible asynchronous spanning tree coverage

A possible spanning tree coverage starting in node L and obtained by this scenario is shown in Fig. 4.2. Part of the scenario body embraced by rule `repeat` will be self-replicating and moving on the forefronts to new nodes within this branching and top-down network coverage.

- Synchronous breadth-first spanning tree coverage starting from some node may look like follows:

      ```
      hopfirst(current);
      repeat_sync(hopfirst(links(all), nodes(any)))
      ```

  or, as information on nodes can be omitted too:

      ```
      hopfirst(current);
      repeat_sync(hopfirst_links(all))
      ```

  One of possible synchronous tree-like coverage by this scenario, starting in node L, is shown in Fig. 4.3.
  Synchronous network coverage may happen to be more expensive than the asynchronous one as may need providing feedback after termination of each cycle to the starting node, from which the repeating scenario body may be broadcast each time to the new network nodes reached. But this body may also be delivered to each new node on the forefront of this spanning tree coverage, like in the previous asynchronous case, and only forward parallel command may be needed from the starting node to activate another iteration in parallel in all nodes reached (see also Chap. 2 for explanation of similar cases).

- Depth-first spanning tree coverage starting from some node (such a tree allowing the remaining, not accounted, network links from its nodes to connect only with the already passed nodes up the tree, and not between different branches):

      ```
      hopfirst(current);
      ```

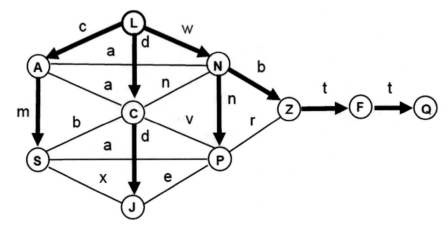

Fig. 4.3 Possible synchronous spanning tree coverage

```
frontal(Depth)  =
    {sequence(hopfirst_links(all); run(Depth))};
run(Depth)
```

This scenario contains procedure Depth used recursively. It automatically splits the first scenario in the two, separated by semicolon (as a hop in each reached node to all new neighboring nodes not visited yet), and then creates branches starting in these reached nodes, from which the second scenario-operand, as same procedure Depth again, continues. These branches will be developing strictly sequentially (using rule sequence), with the next branch starting after full completion of the previous branch (which may itself have hierarchical applications of Depth, and as deep as possible).

One of possible depth-first trees, starting in node L and created by this scenario, is shown in Fig. 4.4, with remaining links as dashed arrows leading only to the already passed nodes up the tree. The body of procedure Depth, as content of same named frontal variable, will be propagating on the forefronts during this top-down coverage of the network.

4.8 Cycles and Paths Through Nodes

Only few from a variety of possible definitions and solutions for such tasks are considered below.

- Cycle through all nodes, starting in the current node, collecting the nodes passed, and providing output in the starting node

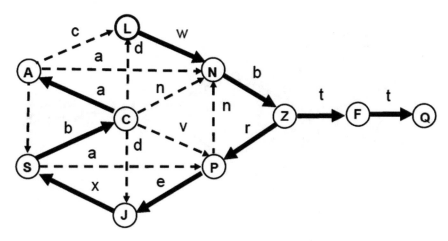

Fig. 4.4 Depth-first spanning tree coverage example

```
frontal(Path); nodal(Mark, Parent);
Path = NAME; Mark = 1;
repeat(
    repeat(or(hop_links(all); empty(Mark); Mark = 1;
                Path &&= NAME; Parent = PREVIOUS)));
    repeat(hop(Parent); Path && = NAME;
            no(hop_link(all); Mark == nil);
            if(NAME == Path[1], quit_output(Path))))
```

Where we used shortened version for

```
Path = append(Path, NAME)   as: Path && = NAME
```

This scenario combines two parts regularly invoked and succeeding each another. First part, going down through only one of the remaining links from the current node, is leading to new nodes as far as possible, i.e. to maximum depth. The second part then returns through the tracks, obtained by the first part, to the first node above (previous nodes recorded in nodal variables `Parent`), which is still having not unvisited neighbours. After this the first part begins working again, and so on, until the starting node is reached.

The forth and back path fragments are collected by the scenario parts into the united path in frontal variable `Path`, which is issued in the scenario starting node. Result in the starting node C may be one of the following (also as shown in Fig. 4.5):

```
(C, N, Z, F, Q, F, Z, P, J, S, A, L, C)
```

- Path through all nodes, with output in the starting node can be found as follows:

```
frontal(Path, Return); nodal(Mark, Parent);
```

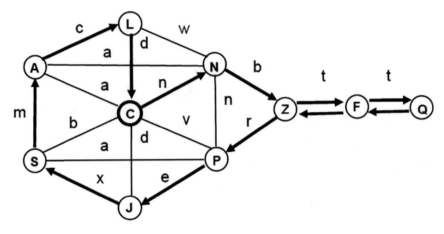

Fig. 4.5 Cycle through all nodes

```
Path = NAME; Mark = 1;
repeat(
   repeat(or(hop_link(all); empty(Mark); Mark = 1;
              Path && = NAME; Parent = PREVIOUS));
   repeat(hop(Parent); Return && = NAME;
          not(hop_link(all); Mark == nil);
          if(NAME == Path[1], quit_output(Path)));
   Path && = Return; Return = nil)
```

This scenario is based on the previous case but is without the backward path segment from the last forwardly reached node to the starting node. For this, the backward fragment of the growing path is first accumulated in a separate frontal variable `Return`, content of which is added to the full path in `Path` except for the last fragment leading via the already visited nodes to the scenario starting node.

The printed result, starting in node C (one of many possible), will be as follows:

```
(C, N, Z, F, Q, F, Z, P, J, S, A, L)
```

- All possible simple paths (i.e. without repeating nodes) from a current node, let it be A in Fig. 4.1, to some node P can be found by:

```
frontal(Path);
repeat(notbelong(NAME, Path); Path && = NAME;
       if(NAME == P, quit_output(Path));
       hop_links(all))
```

Results printed when staying in the final node P:

```
(A,L,N,P), (A,N,P), (A,C,P), (A,S,P), (A,S,J,P),
(A,N,Z,P), (A,N,C,S,P), (A,S,C,L,N,P),
(A,S,J,C,L,N,P), (A,S,J,C,L,N,Z,P)
```

All solutions above use free scenario code movement between nodes during the path accumulation.

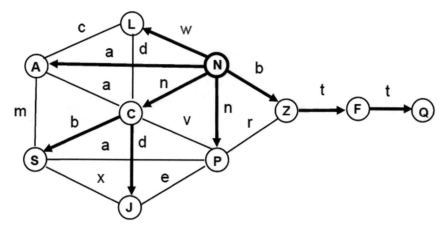

Fig. 4.6 Shortest path tree from node N

4.9   Shortest Path Trees and Shortest Paths

- Starting here from Shortest Path Tree (SPT) creation and registration.

Let us create SPT from node N of Fig. 4.1. In spanning trees coverage shown before we had only processes of network navigation, but in the following solutions the coverage will also be combined with recording such trees within the network structure, for a further use.

Synchronous solution. With all link weights assumed to be the same, we can obtain the SPT from node N in the simplest way synchronously, with the tree registration provided by nodal variables Parent in each node naming the above node in the tree, with one of possible results shown in Fig. 4.6.

```
hopfirst_node(N); nodal(Parent);
repeat_sync(hopfirst_links(all); Parent = PREVIOUS)
```

Asynchronous solution. The most general solution, fully asynchronous and with certain weights associated with links, which can be accessed by environmental variable LINK after passing the link and arriving into node (all these weights may be just 1 in the simplest case) will be as follows.

```
hop_node(N);
nodal(Parent, Distance); frontal(Length);
Distance = 0;
repeat(
   hop_links(all); Length += LINK;
   or(Distance == nil, Distance > Length);
   Distance = Length; Parent = PREVIOUS)
```

- Finding shortest path (SP) on the basis of SPT found. There can be different organizational schemes for doing this.

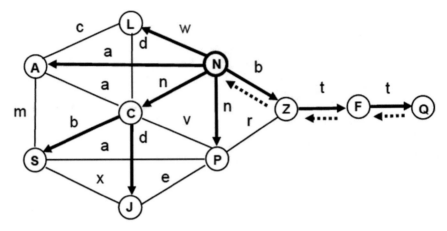

Fig. 4.7 Finding shortest path to node Q after SPT from node N created

(a) *Starting in the final node and issuing the path in the starting node.* The following
    scenario for a path from node N to node Q starts in the final node Q and then
    moves upwards the SPT while collecting the passed path in reverse order unless
    reaches node N, with issuing it there. This scenario should be activated after the
    SPT creation and within the same overall scenario as common variable `Parent`
    is used in both cases (see also Fig. 4.7 showing the path found).

```
hop_node(Q); frontal(Path);
repeat(Path = NAME && Path; hop(Parent);
       if(NAME == N, quit_output(Path)))
```

This scenario may even be simpler, as the cycle up the tree will automatically
terminate in node N as there was no variable `Parent` set up there, so the hop by it
will terminate the spatial cycle in any case:

```
hop_direct(Q); frontal(Path);
repeat(Path = NAME && Path; hop(Parent));
output(Path)
```

Result printed in both case will be as:

```
(N, Z, F, Q)
```

(b) Another solution for the shortest path from node N to node Q, after SPT from
    N created, may be by starting from N (initially staying in it after SPT creation)
    and moving downward the tree and by its full breadth (with accumulating all
    paths passed which may be done in parallel) while providing output in Q of the
    particular path needed upon reaching this node, as follows.

```
frontal(Path);
repeat(Path = append(Path, NAME);
       if(NAME == Q, quit_output(Path)));
```

```
hop_links(all); Parent == PREVIOUS)
```

But this solution may be less efficient than the obtained by previous scenario as regard both time and distributed resources used. Its main advantage may be that the SPT creation and path finding can start in the same node, without the need of direct hopping to final node Q, which may not be easy in complex distributed and unknown environments. The disadvantage—that the final result is output in node Q, not in the starting one N, which may be inconvenient and inefficient too.

This solution, similar to previous one, should also be activated right after SPT creation, in the same scenario, as common variable `Parent` is used for both. And such solution, in order to reduce the load on networking resources used, can also use rule `abort` instead of `quit`, which will trigger cancellation and removal of all other, possibly still continuing, path collecting activities in the network after the needed path is issued.

(c) Beginning in starting node and issuing the result in the starting node too. Shortest path collection from the starting node N to final node Q, after SPT from N created, may also be isued in N by using internal interpretation track mechanisms for bringing back the final remote result obtained in Q, as follows after proper modification of the previous scenario:

```
frontal(Path);

output_repeat(Path = append(Path, NAME);

              if(NAME == Q, blind(Path));

              hop_links(all); Parent == PREVIOUS)
```

(d) Combination of the latest path-collecting solution with asynchronous finding of SPT, both starting from node N within the same, integrated, scenario can be as follows:

```
hop_node(N);

nodal(Parent, Distance); frontal(Length);

Distance = 0;

stay_repeat(hop_links(all); Length += LINK;

            or(Distance == nil, Distance > Length);

            Distance = Length; Parent = PREVIOUS);

frontal(Path);

output_repeat(Path = append(Path, NAME);

              if(NAME == Q, blind(Path));

              hop_links(all); Parent == PREVIOUS)
```

4.10 Network Creation

There may be a variety of options for creating distributed knowledge networks with
any topologies by space navigating parallel scenarios in SGL reflecting proper matching patterns. Some elementary examples are discussed below.

- Creating isolated node Peter and hopping into it:

```
create(hop_node('Peter'))
```

  or, as scenario control will always hop into the node created, on default, this may
be shortened to:

```
create_node('Peter')
```

  or, shorter, also if Peter does not intersect with other language constructs:

```
create(#Peter)
```

  or, as only node can be mentioned here with a single name, this may even be:

```
create(Peter)
```

- Extending this single-node network by a statement like Peter is father
of Alex after initially directly hopping into Peter may be as follows:

```
hop_direct('Peter');
create(hop(link(+'fatherof'), node('Alex')))
```

  or simplified, also if link and node names do not intersect with other SGL constructs:

```
hop(Peter); create(link(+fatherof), node(Alex))
```

  or, as shortest:

```
#Peter;create(+fatherof#Alex)
```

  If node Alex already exists, and only new link to it should be created, we may
write, also in short:

```
#Peter;linkup(+fatherof#Alex)
```

- Creating the whole network.

  We will show this from scratch for the network of Fig. 4.1 by using its depth-
first spanning tree (one such tree, starting from node L, is shown in Fig. 4.4, which
will be followed in the current example). As all links not included into the tree are
connecting current nodes with the nodes passed before and lying on the same paths
up the tree, we can effectively use frontal variables remembering addresses of nodes
to be linked back.

  These nodes will not be created again as they are represented by their addresses
which automatically confirm their existence. The following scenario will create the

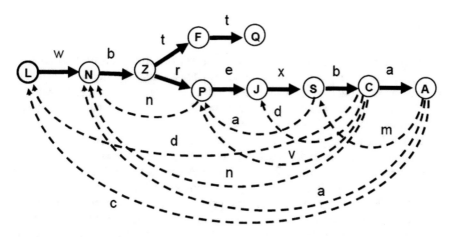

Fig. 4.8 Network creation based on depth-first spanning tree

whole network, with forward and feedback steps depicted in Fig. 4.8 by the use, respectively, of solid and dashed arrows.

```
frontal(FL, FN, FP, FJ, FS);

create(

    #L; FL = ADDR; w#N; FN = ADDR; b#Z; (t#F; t#Q),

    (r#P; FP = ADDR; free(n#FN); e#J; FJ = ADDR; x#S;

    free(a#FP); FS = ADDR; b#C; free(d#FL, n#FN,

    v#FP; d#FJ); a#A; c#FL, a#FN, m#FS))
```

We may also delegate hopping back to the existing nodes entirely to the SGL internal tracking mechanisms, using for this symbol ## as a hop to the already created and passed nodes lying up the tree (similar to SGL rule linkup with modifier passed), so now only new links to them will be created, as follows.

```
create(
    #L; w#N; b#Z; (t#F; t#Q),
    (r#P; free(n##N); e#J; x#S; free(a##P);
    b#C; free(d##L, n##N, v##P; d##J);
    a#A; c##L, a##N, m##S))
```

More details on how to use SGL or its predecessor variants for creation, modification, and processing of distributed networks can be found in many previous publications [1–3, 15–23].

4.11 Conclusion

We have presented some traditional graph and network processing routines in SGL, with their different representation options and possible simplifications, which will be used in the subsequent chapters as basic ones for solving more complex networking problems, which showed their transparency and compactness under the network processing model and technology developed.

SGT, in line with other capabilities for distributed systems applications, can also be considered as a universal spatial machine for parallel and fully distributed processing of any graphs and networks. It allows us to directly land on the network body, move on and through it, and solve any problems including network creation from scratch, its modification, and removal. It supposes using any computational resources accessible from the network nodes via communicating copies of the SGL interpreter distributed throughout physical spaces and implanted into different systems, social ones including, and these resources can be engaged cooperatively and massively. The network solutions under SGT are based on high-level holistic scenarios reflecting active matching patterns, which are freely propagating through the network bodies while carrying proper operations and control, also capable of being replicated and self-modified in this spatial navigation process.

The network processing scenarios are often very simple and compact as they just reflect our mental style of scanning and perception of distributed spaces, with feeling of personal presence in the latter (see also Chap. 2, and more on this will be in Chap. 8). Many traditional technical routines common in other models and languages can be effectively omitted under SGT and delegated to the intelligent networked SGL interpretation.

Other results on parallel and distributed processing of general graphs and networks based on the internal network activity, wavelike navigation and coverage of distributed networks, and the developed spatial pattern matching techniques can be found in [16–23].

References

1. P. Sapaty, *Managing Distributed Dynamic Systems with Spatial Grasp Technology* (Springer, 2017)
2. P. Sapaty, *Ruling Distributed Dynamic Worlds* (Wiley, New York, 2005)
3. P. Sapaty, *Mobile Processing in Distributed and Open Environments* (Wiley, New York, 1999)
4. E.J.H. Chang, Echo algorithms: depth parallel operations on general graphs. IEEE Trans. Softw. Eng. **SE-8**(4), 391–401 (1982)
5. R.J. Wilson, *Introduction to Graph Theory*, 5th edn (Prentice Hall, 2012). https://www.amazo n.com/Introduction-Graph-Theory-Robin-Wilson/dp/027372889X
6. K. Ruohonen, *Graph Theory* (2013). http://math.tut.fi/~ruohonen/GT_English.pdf
7. P.V. Dooren, *Graph Theory and Applications* (Dublin, 2009). http://www.hamilton.ie/ollie/D ownloads/Graph.pdf

8. D. Krackhardt, Assessing the political landscape: structure, cognition, and power in organizations. Adm. Sci. Q. **35**(2), 342–369 (1990). https://www.jstor.org/stable/2393394?origin=cros sref&seq=1#page_scan_tab_contents
9. Krackhardt kite graph. https://en.wikipedia.org/wiki/Krackhardt_kite_graph
10. R. Ackland, *Social Network Analysis*. School of Sociology Centre for Social Research & Methods Australian National University, SOCR8005 Social Science of the Internet, S1 (2016). http://vosonlab.net/papers/ACSPRISummer2017/Lecture_SocialNetworkAnalysis.pdf
11. P. Tubaro, *Introduction to Social Network Analysis* (University of Greenwich, London, 2012). https://paolatubaro.files.wordpress.com/2012/07/workshopbristol_27june2012_v2.pdf
12. R. Moeller, Multimedia information extraction and retrieval. Soc. Netw. Analy. Hamburg University of Technology. https://pdfs.semanticscholar.org/presentation/96ec/b58fce68ea32d283 dcee9018ddb7d85b3f40.pdf
13. V. Krebs, in *Social Network Analysis: an Introduction*. http://www.orgnet.com/sna.html
14. W. Fan, in *Graph pattern matching revised for social network analysis*, ICDT 2012, 26–30 March 2012 (Berlin, Germany). http://homepages.inf.ed.ac.uk/wenfei/papers/icdt12.pdf
15. P.M. Borst, M.J. Corbin, P.S. Sapaty, in *Wave processing of networks and distributed simulation*, Proceeding of HPDC-3 International Conference, San Francisco, Aug 1994, IEEE, (1994), pp. 61–69
16. P.S. Sapaty, WAVE-1: a new ideology of parallel processing on graphs and networks. Future Generations Comput. Syst. **4** (1988). North-Holland
17. P.S. Sapaty, The WAVE-1: a new ideology and language of distributed processing on graphs and networks. Comput. Artif. Intell. (5) (1987)
18. P.S. Sapaty, A wave language for parallel processing of semantic networks. Comput. Artif. Intell. **5**(4) (1986)
19. P.S. Sapaty, in *The wave approach to distributed processing of graphs and networks*, Proceedings of the International Working Conference on Knowledge and Vision Processing Systems, Smolenice, Nov 1986
20. P.S. Sapaty, in *Active information field as a model for structural solving of tasks on graphs and networks*, Proceeding of USSR Academy of Sciences. Technical Cybernetics, 1984 (in Russian)
21. P.S. Sapaty, in *Solving tasks on semantic networks and graphs by active distributed structures*, Proceeding of 3rd International Conference Artificial Intelligence and Information-Control Systems of Robots, Smolenice, Elsevier Science Publishers B.V., North-Holland, 1984
22. P.S. Sapaty, in *On efficient structural implementation of operations on semantic networks*, Proceeding of USSR Academy of Sciences. Technical Cybernetics, issue 5 (1983) (in Russian)
23. P.S. Sapaty, in *A structural approach to solving tasks of the graph and network theory*, Proceedings of 5th All-Union School on Parallel Programming and High-Performance Systems, Kiev, Naukova Dumka, 1982 (in Russian)

Chapter 5
Social Networks Processing Under SGT

5.1 Introduction

Detailed solutions for discovering and analysing basic and widely recognized features and parameters of social networks will be presented and explained in SGL. These will include, first of all, the centrality issues like degree centrality, eigenvector centrality, closeness centrality, and betweenness centrality of its nodes. Also considered will be network clustering, finding both strongest subnetworks (or cliques) and weakest (or articulation points), with possible network topology changes for properly dealing with them for different applications.

Other solutions will be for recognizing particular structures using arbitrary complex search and matching patterns which may contain variants and alternatives. Will also be considered how to use physical parameters of social networks if node coordinates are available, like calculating whole regions occupied by the network, finding locations of topological centers of different communities and distances between them, which may be useful for predicting and preventing possible conflicts between such communities. A possibility of modelling social network dynamics in physical spaces will be shown in SGL too.

All demonstrated scenarios can be using arbitrary large networks without any borders, where nodes and links may be arbitrarily distributed in physical spaces, and no centralized resources needed for solving complex problems.

The solutions shown in this chapter are based on previous applications of the distributed pattern-matching approach for solving network problems, summarized in the previous three books [1–3]. They are also inspired by existing publications in this area, with [4–11] as some of them.

© Springer Nature Switzerland AG 2019
P. S. Sapaty, *Holistic Analysis and Management of Distributed Social Systems*, Studies in Systems, Decision and Control 184, https://doi.org/10.1007/978-3-030-01830-6_5

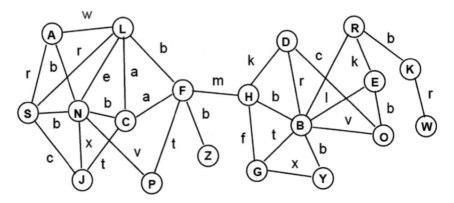

Fig. 5.1 Exemplary social network structure

5.2 Centrality Issues

One of the key applications in social networks is to identify the most important, or central, nodes in the network. The measure of centrality is used to give a rough indication of the social power nodes based on how well they connect the network. In social network analysis, Degree, Betweenness, and Closeness centrality [6–11] are three most popular methods to measure the centrality in a social network.

For convenience, in the following analysis we will be using the network topology example shown in Fig. 5.1 which is often exhibited in other publications considering the centrality issues [7–10], but with our own node and link names.

5.2.1 Degree Centrality

Degree centrality is defined as the number of edges incident upon a node, being usually the first way to calculate the nodes that are most potential in comparison with other nodes. If the edges in a graph are directed, the *in-degree* centrality is differentiated from the *out-degree* centrality.

To determine the degree centrality of node A we should write in SGL:

```
hop_node(A); output_count(hop_links(all))
```

This degree, as 3, will be issued in node A when staying in it (degree centrality for all network nodes are shown in Fig. 5.2).

And to find the node with maximum degree and output its name prefixed with this degree we may write (see also related examples in Chap. 4):

```
output_max(hop_node(all);
           count_hop_link(all) && NAME)
```

which will give for our network as:

```
(7, B)
```

Or if to output node's name only, the scenario modification may be as follows:

```
output_element(max(hop_node(all);
                count_hop_link(all) && NAME), 2)
```

which will just give:

B

5.2.2 Eigenvector Centrality

In graph theory, *eigenvector centrality* (also called *eigencentrality*) is a measure of the influence of a node in a network. It assigns relative scores to all nodes in the network based on the concept that connections to high-scoring nodes contribute more to the score of the node in question than equal connections to low-scoring nodes.

We may initially accumulate each node centrality degree, as the number of its links to other nodes, in nodal variable `Degree` associated with each node, as follows:

```
nodal(Degree); hop_nodes(all);
Degree = count_hop_links(all))
```

The values in `Degree` in each node will correspond to what is shown in Fig. 5.2.

In the first step toward calculating eigenvector centrality, we may substitute the value of `Degree` in each node by the sum of obtained values in variables `Degree` in all its neighbouring nodes, thus taking into account the importance of each node by the values of its neighbours, as follows:

```
nodal(Degree);
sequence(
  (hop_nodes(all); Degree = count_hop_links(all)),
```

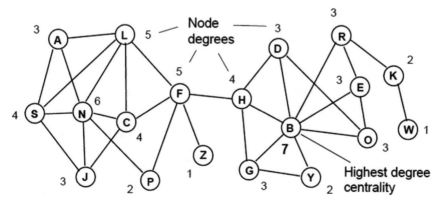

Fig. 5.2 Degree centrality of nodes

```
(hop_nodes(all);
Degree = sum(hop_links(all); Degree)))
```

Highest in eigenvector centrality after this first step of taking into account weights of neighbours is node L (having now weight 22). Nodes N and B are the closest to L in this estimate (both having 21), see Fig. 5.3.

To increase the influence of not only direct neighbours on the node's value, but also neighbours to neighbours, and so on, we may organize the second part of the previous scenario as operating repetitively and synchronously, with some threshold number of repetitions given in some nodal variable Steps, as follows:

```
nodal(Degree, Steps = number);
sequence(
   (hop_nodes(all); Degree = count_hop_links(all)),
   repeat_sync_Steps(
      hop_nodes(all);
      Degree = sum(hop_links(all); Degree)))
```

This repetitive spreading of influence of more powerful nodes may be changing the importance of nodes in relation to each other. For example, after the second step of accounting weights of neighbours (with 2 given in Steps), node N (with 94) will dominate, as in Fig. 5.4.

With the third step, the dominance of node N continues to grow (now weighing 367), as in Fig. 5.5.

The subsequent iterations will see further increase of the weight of N, despite it having lower than node B level of degree centrality, as in Fig. 5.2 (i.e. 6 against 7).

Another example of rapidly growing and spreading weights of nodes within higher connected zones in the network is shown in Fig. 5.6, with initial degree centrality shown at all nodes.

The first step in eigenvector centrality will result in node K as the highest one (12), as in Fig. 5.7.

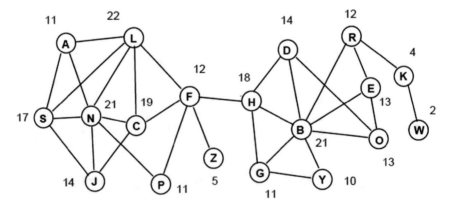

Fig. 5.3 Using weights of neighbouring nodes, first step in eigenvector centrality

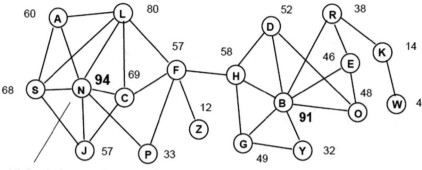

Highest eigenvector centrality

Fig. 5.4 Using weights of neighbouring nodes, second step in eigenvector centrality

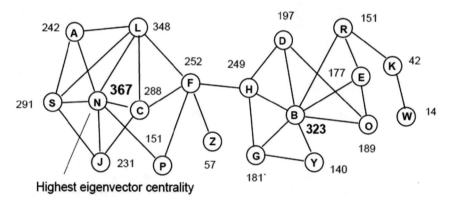

Highest eigenvector centrality

Fig. 5.5 Third step in eigenvector centrality

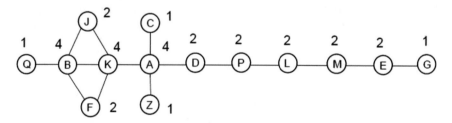

Fig. 5.6 Another network example with degree centrality

In further iterations, all nodes of the more interconnected left side of the network are growing rapidly in value (with node K remaining the leader) in comparison with the right part of the network, with the results of the fourth iteration shown in Fig. 5.8.

All considered eigenvector examples were based on synchronous network access for each iteration, where every step started from activating all network nodes again.

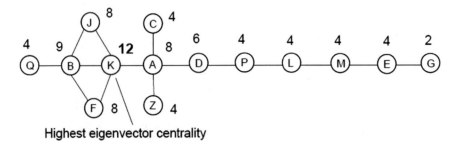

Highest eigenvector centrality

Fig. 5.7 First step of calculating eigenvector centrality

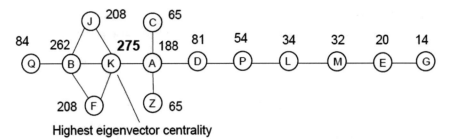

Highest eigenvector centrality

Fig. 5.8 Fourth step of calculating eigenvector centrality

Using asynchronous repetition where all nodes are simultaneously accessed only at the beginning, and the subsequent iterations all performed by direct communication between neighboring nodes without global synchronization, may somehow alter intermediate results, but the global tendency will persist. The asynchronous solution, with limiting the number of such eigenvector iterations in each node and providing the output of the strongest node at the end with its eigenvector value, may be as follows (also providing some time delay in nodes before the next local iteration, using nodal variable Delay).

```
nodal(Degree, Steps = number, Delay = time);
output_max(
align(hop_nodes(all);
     Degree = count_hop_links(all));
repeat_Steps(Degree = sum(hop_links(all); Degree);
            sleep(Delay));
Degree && NAME)
```

This might also result, as before, in (275, K) for the network of Fig. 5.8 if cycles in all network nodes operated in parallel and Steps was set up to 4. The established Delay in each node was introduced for possible balancing of the distributed asynchronous interlinked operations, especially in making them closer in time of independent iterations in different nodes.

5.2.3 Closeness Centrality

The measure of closeness centrality is to take into account how distant a node is to the other nodes in the network. The closeness centrality for a node may be evaluated by the sum of all shortest paths from it to all other network nodes, or by mean, or average, shortest path to other nodes. (The latter measure may sometimes happen to be more convenient, especially when comparing results for nodes from different networks with quite different sizes.)

(a) Finding the sum of shortest paths from some node A of Fig. 5.1 to all other nodes after shortest path tree (SPT) created from A.

We may use the SPT creation scenario from Chap. 4 where shortest distance in each node of the tree from the root, to be A in our case, is recorded in nodal variable Distance. For finding the sum of all shortest paths from node A after SPT from A has been created we may write:

```
hop_node(A);
output_sum(hop_nodes(all_other); Distance)
```

or for the average of all shortest paths from A:

```
hop_node(A);
output_average(hop_nodes(all_other); Distance)
```

The full solution, for example, for the sum of all shortest paths from A, which includes the asynchronous SPT creation scenario borrowed from Chap. 4 (without fixing predecessor nodes in the tree as redundant here), will look like follows (taking for simplicity all link weights as 1):

```
hop_node(A);
nodal(Distance); frontal(Length);
stay(
  Distance = 0; Length = 0;
  repeat(
    hop_links(all); Length += 1;
    or(Distance == nil, Distance > Length);
    Distance = Length));
output_sum(hop_nodes(all_other); Distance)
```

(b) We can easily extend this scenario for simultaneous finding of sums of shortest paths from all network nodes with subsequent choosing the node with minimum such sum, to be named as highest closeness centrality node, as follows:

```
output_min(
  hop_nodes(all); IDENTITY = ADDRESS;
  nodal(Distance); frontal(Length);
  stay(
    Distance = 0; Length = 0;
    repeat(
      hop_links(all); Length += 1;
      or(Distance == nil, Distance > Length);
```

```
Distance = Length));
    append(sum(hop_nodes(all_other); Distance), NAME))
```

For example, for node F we will have the following summation of its shortest paths to all other nodes:

```
1 + 1+1 + 1+1 + 2+2 + 2+2 + 2+2 + 2+3 + 3+3 + 3+4 + 5  =  40
```

Whereas for node H:

```
1 + 1+1 + 1+2 + 2+2 + 2+2 + 2+2 + 2+3 + 3+3 + 3+3 + 4  =  39
```

And for node D:

```
1 + 1+1 + 2+2 + 2+2 + 2+2 + 3+3 + 3+3 + 3+4 + 4+4 + 4  =  42
```

After considering all remaining nodes too, we will find node H as the closest one to all other nodes, see Fig. 5.9. So the printed result will be as:

```
(39, H)
```

5.2.4 Betweenness Centrality

Betweenness centrality is another key metrics for computing the importance of nodes, as an assessment of how a node lies between other nodes in the network. If a node is the only node that links two groups of nodes in the network, like nodes F and H in the previous figures (nodes R and K formally too, but with much weaker positions), this node shall be seen as an important one for keeping the social network together. Therefore, betweenness centrality is to measure the connectivity of the neighbours of a node and to give a higher value for nodes which bridge clusters.

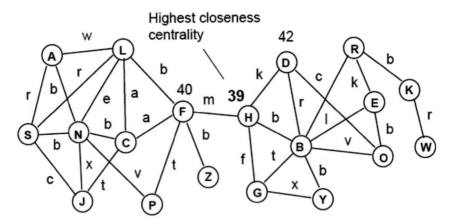

Fig. 5.9 Closeness centrality of nodes

In graph theory, *betweenness centrality* is a measure of centrality in a graph based on shortest paths. For every pair of vertices in a connected graph, there exists at least one shortest path between the vertices such that either the number of edges that the path passes through (for unweighted graphs) or the sum of the weights of the edges (for weighted graphs) is minimized. The betweenness centrality for each vertex is *the number of these shortest paths* that pass through the vertex.

Let us consider stepwise organization of the solution for betweenness centrality using as an example the network topology of Fig. 5.1.

(a) After SPT is supposedly created from node A to all other nodes, as in Chap. 4 (with the use of its nodal variables `Parent`), find how many of the shortest paths (SPs) from A to all other nodes are passing via node D, with output of the received betweenness centrality measure in node D.

```
hop_node(D); nodal(Between);
stay(
  hop_node(A);
  hop_nodes(all_other); nonequal(NAME, D);
  repeat(hop(Parent);
         if(NAME == D, blind_incremet(Between)))));
output(Between)
```

(b) Full solution including SPT creation from node A will be as follows:

```
hop_node(D); nodal(Between));
stay(
  hop_node(A);
  nodal(Parent, Distance); frontal(Length);
  stay_repeat(
       hop_links(all); incremenmt(Length);
       or(Distance == nil; Distance > Length);
       Distance = Length; Parent = PREVIOUS);
  hop_nodes(all_other); nonequal(NAME, D);
  repeat(hop(Parent);
         if(NAME == D, blind_incremet(Between)))));
output(Between)
```

(c) Now, for all SPs between all couples of nodes (allowing passing an SP between two nodes only in one direction, by comparing addresses of the starting and final nodes), and for particular node D as its betweenness centrality value we will have:

```
hop_node(D); nodal(Between));
stay(
  hop_nodes(all_other); IDENTITY = NAME;
  nodal(Parent, Distance); frontal(Length);
  stay_repeat(
       hop_link(all); incremenmt(Length);
       or(Distance == nil; Distance > Length);
       Distance = Length; Parent = PREVIOUS);
  hop_nodes(all_other); nonequal(NAME, D);
  PREVOIUS > ADDRESS;
```

```
      repeat(hop(Parent);
            if(NAME == D, blind_incremet(Between)))));
output(Between)
```

(d) Next and final step: for all SPs between all pairs of nodes, and for all nodes lying inside (i.e. not at the ends) of all these SPs, with determining the node with maximum number of SPs passing through it. We may start from any node or from the network outside, as follows.

```
output_max(
  nodal(Between);
  hop_nodes(all); IDENTITY = NAME;
  nodal(Parent, Distance); frontal(Length);
  stay_repeat(
    hop_link(all); incremenmt(Length);
    or(Distance == nil; Distance > Length);
    Distance = Length; Parent = PREVIOUS);
  stay(hop_nodes(all_other); PREVIOUS > ADDRESS;
      repeat(hop(Parent);
              if(nonequal(IDENTITY, NAME),
                incremet(Between)))));
append(Between, NAME))
```

Nodal variables `Between` will be having global, default, identity same for all nodes, whereas nodal variables `Parent` and `Distance` will be preserving identities of individual nodes the SPTs are starting from (with these trees spatially intersecting but not interfering with each other due to this).

The output by this scenario for the network considered will be:

```
(87, F)
```

It means the highest betweenness centrality will be for node `F`, as shown in Fig. 5.10.

The number of shortest paths via node `F` will be: `87`.

Via seemingly competitive node `H` this number will be `81`.

5.3 Clustering

Many social networks contain subsets of nodes that are highly connected within the subset and have relatively few connections to nodes outside the subset. The nodes in such subsets are likely to share some attributes and form their own communities. Since the detection of these community structures is not trivial, how to efficiently and effectively discover such community structures is important.

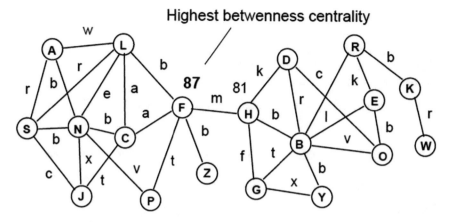

Highest betwenness centrality

Fig. 5.10 Betweenness centrality

5.3.1 Clustering Coefficient and Examples of Clusters

A clustering coefficient is to measure the degrees of nodes to decide which nodes in a graph tend to be clustered together. Thus, the clustering coefficient measure is often to quantify how close its neighbours are to a complete graph.

We will try here only very rough estimates for clustering, analysing network structures and not taking into account the semantics of nodes and inter-node relations, which in reality may be dominant for the clustering. We will consider outlining possible clusters starting from some node and covering node's vicinity to a certain depth, with using a coefficient reflecting intensity of inter-node relations in this group (by comparing the number of existing links between cluster nodes to their maximum possible number as in full graph, or clique), as follows.

```
nodal(Start = nodename, Cluster, Number,
      Maxlinks, Linksintern, Coefficient);
frontal(Depth = depth);
hopfirst_node(Start);
Cluster = NAME &&
   repeat_Depth(hopfist_links(all); free(NAME));
Number = count(Cluster);
Maxlinks = (Number - 1) * Number / 2;
Linksintern =
   count(hop_nodes(Cluster); hop(links(all),
         nodes(Cluster))) / 2;
Coefficient = Linksintern / Maxlinks;
output(Coefficient, Cluster)
```

Starting in node A with Depth == 1, we will have the cluster as shown in bold in Fig. 5.11, which appears to be a full clique (congratulating ourselves with the success!).

The scenario provides the following parameters for this Depth:

- Nodes belonging to the cluster: `Cluster == (A, L, S, N)`
- Number of cluster nodes: `Number == 4`
- Maximum possible number of links with this number nodes: `Maxlinks == 6`
- Internal cluster links number: `Linksintern == 6`
- Clustering coefficient: `Coefficient == 1`

Printed result in the point we started the cluster, i.e. node A:

```
1, (A, L, S, N)
```

If to start in node A with `Depth == 2`, we will be having the situation shown in Fig. 5.12.

With the following parameters obtained for this case:

- Nodes belonging to the cluster: `Cluster == (A, L, S, N, J, C, P, F)`
- Number of cluster nodes: `Number == 8`
- Maximum possible number of links with these nodes: `Maxlinks == 28`
- Internal cluster links number: `Linksintern == 15`
- Clustering coefficient: `Coefficient == 0.54`

The printed result when staying in node A:

```
0.54, (A, L, S, N, J, C, P, F)
```

5.3.2   Importance of Inside to Outside Links Ratio

Except the used above clustering coefficient showing how close the supposed cluster is to the full graph for the nodes chosen, another parameter may be very important too, like the ratio between the number of links inside the cluster and those leading outside, for which we may extend the scenario above as follows.

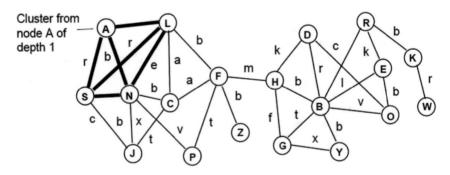

Fig. 5.11 A cluster from A of depth 1 as a clique

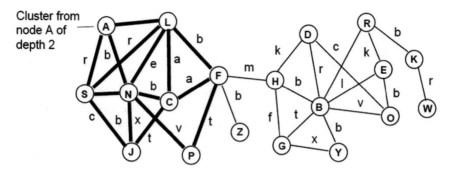

Fig. 5.12 A cluster from node A with depth 2

```
nodal(Start = nodename, Number, Maxlinks,
        Linksintern, Linksextern, Coefficient,
        Inoutratio);
frontal(Cluster, Depth = depth);
hopfirst_node(Start);
Cluster = NAME &&
    repeat_Depth(hopfist_links(all); free(NAME));
Number = count(Cluster);
Maxlinks = (Number - 1) * Number / 2;
Linksintern =
    count(hop_nodes(Cluster);
            hop(links(all), nodes(Cluster))) / 2;
Linksextern = count(
    hop_nodes(Cluster);
    hop(links(all), notbelong_nodes(Cluster)));
Coefficient = Linksintern / Maxlinks;
Inoutratio = Linksintern / Linksextern;
Output(Coefficient, Inoutratio, Cluster)
```

With this, we will have the extended previous results for Depths 1 and 2 as follows:

- For Depth 1:

```
Linksintern == 6
Linksextern == 6
Inoutratio == 1
```
Printed result: 1, 1, (A, L, S, N)

- For Depth 2:

```
Linksintern == 15
Linksextern == 2
Inoutratio == 7.5
```
Printed result: `0.54, 7.5, (A, L, S, N, J, C, P, F)`

Despite the first case being full graph, the second case may be preferable to be treated as a cluster as having `7.5` times more internal links than external, whereas in the first case the number of internal links equals the number of external ones.

5.3.3   Using United Clustering Parameter

We may introduce a broader measure of a possible cluster, by combining considered closeness to full graph as `Coefficient` and ratio between its internal and external links as `Inoutratio`—in the form of product of the two, calling the resultant parameter as `United`:

```
United = Coefficient * Inoutratio
```

This will give for the above case of Depth 1:

```
United == 1
```

And for the case of Depth 2:

```
United == 4.05
```

Using this united criteria, we may try to find the best cluster among all possible clusters starting from all nodes of the network with the given `Depth` of network coverage and also some minimum and maximum, or threshold, number of nodes to be considered as potential clusters. The resultant SGL scenario will be as follows:

```
nodal(Cluster, Maxlinks, Linksintern,
      Linksextern, Coefficient;
         Nodesmin = minnumber Nodesmax = maxnumber);
frontal(Depth = depth);
output_max(
  hopfirst_nodes(all);
  Cluster = NAME &&
     repeat_Depth(hopfist_links(all); free(NAME));
  Nodesmin <= (Number = count(Cluster)) <= Nodesmax;
  Maxlinks = (Number - 1) * Number / 2;
  Linksintern =
    count(hop_nodes(Cluster);
           hop(links(all), nodes(Cluster))) / 2;
  Linksextern = count(
       hop_nodes(Cluster);
       hop(links(all), notbelong_nodes(Cluster)));
  United =
    Linksintern ** 2 / (Maxlinks * Linksextern);
  United && Cluster)
```

By this scenario the previous cluster started from node A with Depth == 2 without limiting the number of its nodes will definitely be considered as the best one for this depth, with already mentioned parameters:

```
United == 4.05
Cluster == (A, L, S, N, J, C, P, F)
```

We may try to consider another cluster with Depth 2 (without node limits too), for example, starting from node H (see Fig. 5.13), for which will receive the following results:

```
United == 0.59
Cluster == (H, F, L, C, P, Z, D, B, G, Y, R, E, O)
```

But this cluster will be definitely much weaker than the one started from A.

5.4 Finding Strongest Subnetworks, or Cliques

5.4.1 Finding All Cliques in a Distributed Network

We present here a universal solution in SGL for finding all cliques in a network (i.e. maximum possible full subgraphs), assuming their number of nodes should not be

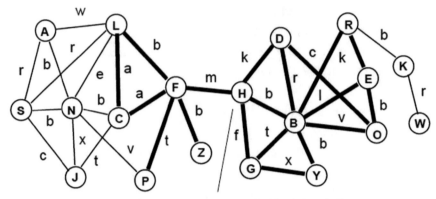

Possible cluster from node H of depth 2

Fig. 5.13 Another possible cluster of depth 2

lower than some threshold. Cliques are of great interest in the graph theory, and they may also have practical importance for social systems as a sort of ideal variant for clustering. The following scenario is finding all cliques in our network in parallel with the number of nodes in them not less than three.

```
hop_nodes(all); frontal(Clique) = NAME;

repeat(hop_links(all); not_belong(NAME, Clique);

        yes(and_parallel(

             hop(links(any), nodes(Clique))));

        if(PREDECESSOR > NAME,

             append(Clique, NAME), blind));

count(Clique) >= 3; output(Clique)
```

This scenario, starting in all nodes and following their links to other nodes in parallel, is collecting node names in new individual hops which have links with all previously collected nodes unless such nodes cannot be found, declaring the collected set of node names in the moving, frontal variable `Clique` as a new clique.

As the same clique can be collected in such a way when started from all nodes of the same clique, the duplicates are blocked above by allowing inclusion of any new node into the clique's list only if the value of its name is lower than of the previous one. Also in the end, by calculating the number of elements in the collected clique's list, the clique is issued only if this number satisfies the given threshold. Many independent branches each trying to collect its own clique will be developing in the distributed network space in parallel.

All cliques will be found in the network considered above (only one four-node, while others are all triangles); they are shown together in Fig. 5.14 (sharing nodes and links with the neighbouring cliques in the same picture).

The flowing cliques will be issued, in the nodes where heir collection terminated:

```
(A, L, N, S),
(J, N, S), (C, J, N), (C, L, N), (C, F, L), (B, D, H),
(B, G, H), (B, G, Y), (B, D, O), (B, E, R), (B, E, O)
```

To output all cliques in the point where the whole scenario started (which may be some network node or outside the network), we may change this scenario to the one as follows:

```
output(
  hop_nodes(all); frontal(Clique) = NAME;
  contain_repeat(
      hop_links(all); not_belong(NAME, Clique);
      yes(and_parallel(
            hop(links(any), nodes(Clique))));
      if(PREDECESSOR > NAME,
         append(Clique, NAME), abort));
  count(Clique) >= 3; unit(Clique))
```

Finding cliques in a network may be important in different situations. For example, these may be very useful for keeping certain infrastructures strong and should be maintained by any means. To speak of adversaries, cliques may be most dangerous groupings in their organizational structures which should be analyzed and dealt accordingly, say, by removal some or all clique links, possibly nodes too, if needed.

5.4.2 Weakening Cliques by Removing Their Links or Nodes

The following scenario after finding cliques removes only all internal clique links for the cliques with the number of nodes not less than 4 (we have only one such clique), leaving nodes which, for example, may continue to be useful for some other network activities, as shown in Fig. 5.15.

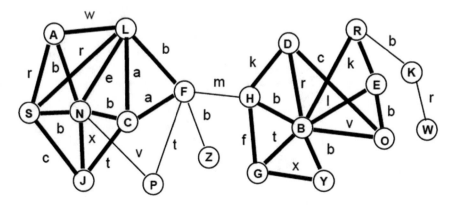

Fig. 5.14 All cliques in the network with the number of nodes not less than three

```
hop_nodes(all); frontal(Clique) = NAME;
repeat(hop_links(all); not_belong(NAME, Clique);
        yes(and_parallel(
                hop(links(any), nodes(Clique))));
        if(PREDECESSOR > NAME,
            append(Clique, NAME), blind));
    count(Clique) >= 4;
    hop_nodes(Clique);
    unlink(links(all), nodes(Clique))
```

As already mentioned, clique nodes may be removed too (with all their links) for a stronger impact on the network, as shown in Fig. 5.16 for the four-node clique considered before.

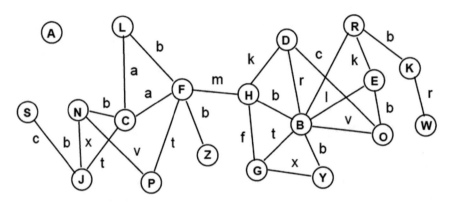

Fig. 5.15 Destroying four-node clique by removing all links between its nodes

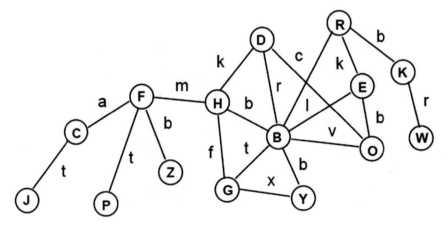

Fig. 5.16 Full removal of the clique with four nodes

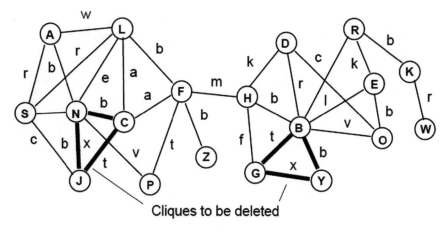

Fig. 5.17 Another, specific, cliques to be found and removed

To accomplish this, the last two strings in the previous scenario should be substituted by:

```
hop_nodes(Clique); delete(NAME)
```

In another example, we may need to find and completely remove (all their nodes and associated links) the cliques with certain links (say, having all three links named x, t, and b between their nodes), which may be classified like "unwanted clusters"). All such cliques, containing only three nodes for our network, are shown in Fig. 5.17. The following scenario will be doing this job.

```
hop_nodes(all);
frontal(Clique = NAME, Links, Concern = (x, t, b));
repeat(
    hop_links(all); not_belong(NAME, Clique);
    yes(and_parallel(hop(links(any), nodes(Clique))));
    if(PREDECESSOR > NAME,
        (append(Clique, NAME); append(Links, LINK)),
        blind));
count(Clique) >= 3;
belong(Concern, Links);
hop_nodes(Clique); delete(NAME)
```

The resultant network freed from these cliques is shown in Fig. 5.18.

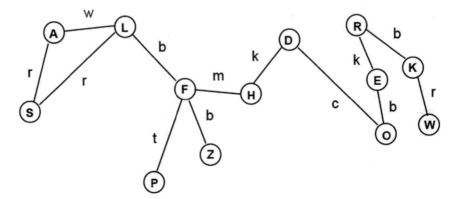

Fig. 5.18 The resultant network after deleting cliques having all links x, t, b

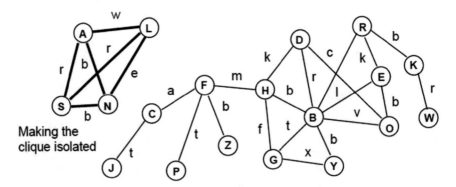

Fig. 5.19 Making the clique fully isolated

5.4.3   Possible Cliques Isolation

Another dealing with cliques, for example, considered as adversary or dangerous ones, may be by their isolation from other network nodes, as shown by the scenario below (which removes all external links for the four node cliques, to be the only one), see also Fig. 5.19.

```
hop_nodes(all); frontal(Clique) = NAME;
repeat(
   hop_links(all); not_belong(NAME, Clique);
   yes(and_parallel(hop(links(any), nodes(Clique)))));
   if(PREDECESSOR > NAME, append(Clique, NAME), blind));
count(Clique) >= 4;
hop_nodes(Clique);
unlink(links(all), notbelong(node, Clique));
```

Leaving clique's nodes and their internal links intact, unlike the previous examples deleting links and nodes, may happen to be useful for a further investigation of their internal organization and activity while blocking external contacts in both ways, from inside and outside. Similar isolation technique can also be used for the cluster examples described earlier and not qualified as cliques.

5.5 Discovering Weakest, or Articulation, Points

Four articulation points F, H, R, K for the network under consideration are highlighted and shown in Fig. 5.20. Each of them, when removed, splits the network into disjoint parts, with first two, F and H, being of highest interest and value. These nodes, for example, may be particularly important for strictest dealing with adversary's systems by just removing (or negotiating with) the most sensitive and vulnerable points (individuals or organizations) in them.

5.5.1 Parallel Finding of All Articulation Points in a Network

Below is parallel and fully distributed solution for finding articulation points in the whole network, by which each network node, having first selected one neighbour randomly, tries to navigate and mark the whole network from it while excluding itself from this process. After termination of the latter, if the node discovers still unmarked neighbours, it declares itself as articulation point and outputs its name, as follows.

```
hop_nodes(all); IDENTITY = NAME;
hopfirst_node(current);
stay(hopfirst_random(links(all)));
    repeat(hopfirst_link(all)));
if(hopfirst_links(all), output(NAME))
```

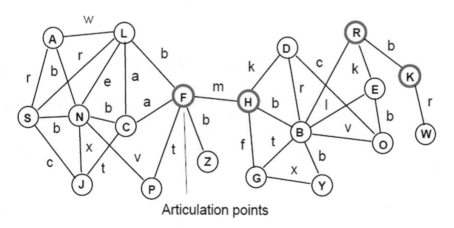

Articulation points

Fig. 5.20 Articulation points of the network

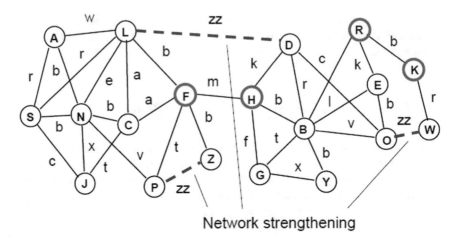

Network strengthening

Fig. 5.21 Strengthening the network by adding new links between nodes

Articulation points will be issued by such nodes themselves, i.e. by F, H, R, and K, which will just output their own names. All articulation node names can also be issued in the starting scenario application node or outside the network by the following scenario variant:

```
output(
    hop_nodes(all); IDENTITY = NAME;
hopfirst_node(current);
    stay(hopfirst_random(links(all)));
        repeat(hopfirst_link(all)));
    if(hopfirst_links(all), NAME))
```

Output in a starting, application node will be:

```
F, H, R, K
```

5.5.2 Strengthening the Network by Adding New Links

After finding articulation points in the network, we may want to strengthen the network, say, by adding new relations between other network nodes, as in Fig. 5.21 (of course, if we have sufficient knowledge about the network topology). This can be done by SGL code that follows (which offers to do this in parallel for the three depicted cases).

```
parallel(
    (hop_direct(L); linkup(zz, D)),
    (hop_direct(P); linkup(zz, Z)),
    (hop_direct(O); linkup(zz, W)))
```

After adding these symbolically zz-named links, the network in Fig. 5.21 will be free from articulation points. Finding optimum solutions for which particular nodes should be additionally interlinked may, however, not be trivial and may need nonlocal analysis and knowledge of (may be even the whole) network structure.

An automatic solution for the network strengthening without any knowledge of the network topology may, however, be possible, as follows. Starting in each node discovered as articulation one, this is to additionally connect its neighbouring node via which the network marking was performed, with all other neighbours that appeared to be unmarked (i.e. unreachable) after termination of the network navigation and marking. The following scenario will be doing this after finding articulation points (the address of the neighbouring node through which network marking was performed is remembered in nodal variable Start, with names of added links as zz again).

```
hop_nodes(all);  IDENTITY  =  NAME;
nodal(Start);  hopfirst_node(current);
stay(hopfirst_random(links(all));
      (hop(BACK);  Start)  =  ADDRESS;
      repeat(hopfirst_links(all)));
hopfirst_links(all);
linkup(zz,  node(hop(BACK);  Start))
```

If randomly chosen neighbouring node from F (with its name remembered in Start) was L, and remaining unmarked nodes after global network navigation via L appeared to be Z and H, the two additional zz links from L to Z and H will be created, as shown in Fig. 5.22.

If randomly chosen neighbouring node from F is H, we will be having another automatic network strengthening variant, as shown in Fig. 5.23.

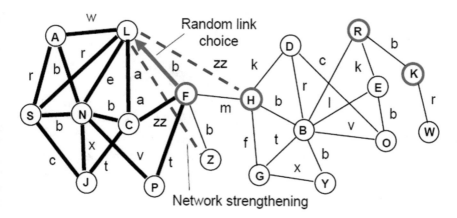

Fig. 5.22 Automatic strengthening the network by adding new links from L to Z and H

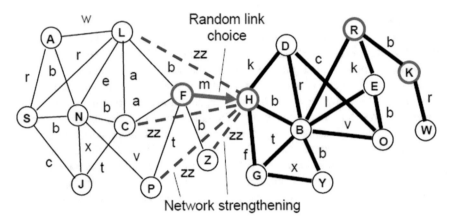

Fig. 5.23 Another automatic network strengthening

5.6   Discovering Particular Structures in a Social Network

5.6.1   A Possible Team Finding Scenario

Let us consider a sort of a possible team (or cluster) structure shown in Fig. 5.24, which may treat the top node as a group's boss/leader having direct links with all group members, and the latter should all communicate with some other group members, at least one, otherwise the group cannot be considered as operational.

The following scenario will be finding all such possible teams, which can be many, but with giving Threshold value on their minimal number of nodes we can restrict the number of such groups found.

Fig. 5.24 Possible organization of a cooperating team

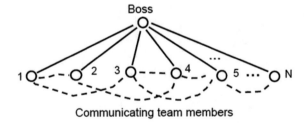

Communicating team members

```
hop_nodes(any); IDENTITY = NAME;
nodal(Cluster, Mark, Threshold = minnumber);
Cluster = NAME &&
   (align(hop_links(all); Mark = 1);
    yes(hop_links(all); nonempty(Mark)); NAME);
count(Cluster) >= Threshold; output(Cluser)
```

The found groups/clusters under the `Threshold` given will be issued in the starting, or "boss", nodes of each cluster. By establishing `Threshold` as 6, we will be getting the following groups:

```
(L,A,S,N,C,F),  (N,S,A,L,C,J)
```

We can easily modify the previous scenario to be capable of finding such a structure with maximum possible number of nodes, as follows:

```
output_max(
   hop_nodes(all); IDENTITY = NAME;
   nodal(Cluster, Mark);
   Cluster = NAME &&
      (align(hop_links(all); Mark = 1);
       yes(hop_links(all); nonempty(Mark)); NAME);
   count(Cluster) && unit(Cluster))
```

The obtained output will be the group of eight nodes with "boss" node as B:

```
8, (B, D, H, G, Y, R, E, O)
```

Two of the mentioned above groups, starting from nodes L and B, are shown in Fig. 5.25.

This result will be issued at the point this scenario was applied, which may be any node of the network or a position outside the network.

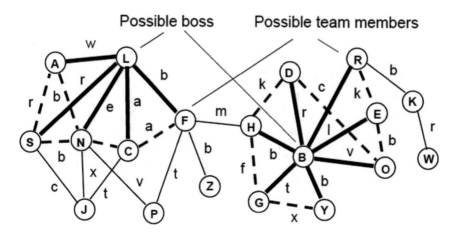

Fig. 5.25 Finding possible team structures in the network

5.6.2   Finding Arbitrary Structures in a Network

It is possible to describe in SGL any graph pattern reflecting any possible situation in distributed social systems (with variables in both nodes and links) and organize its parallel and fully distributed matching with the networked system. An example of such pattern is in Fig. 5.26, with its SGL coding based on a path through all nodes shown below.

```
hop_nodes(all); frontal(Pattern) = NAME;
repeat_4(
    hop_links(all);
    notbelong(NAME, Pattern); Pattern && = NAME);
yes_and(hop(links(all), Pattern[1,2]));
output(Pattern)
```

Two matches of this pattern for the network considered (there are other matches too) are shown in Fig. 5.27.

Some printed matchings, including those of Fig. 5.27, will be as follows:

(A,L,F,C,N), (C,N,S,A,L), (D,B,Y,G,H), (G,H,D,O,B), ...

Another pattern with its representation by a path through all nodes is shown in Fig. 5.28, with SGL coding following.

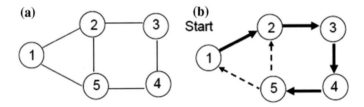

Fig. 5.26   Representing arbitrary graph pattern (**a**) by a path through all nodes (**b**)

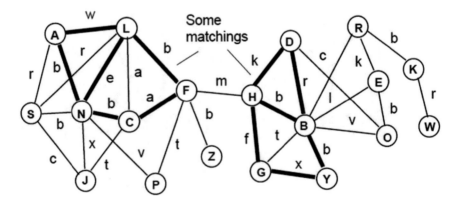

Fig. 5.27   Examples of matching the pattern of Fig. 5.26 with the network

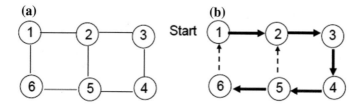

Fig. 5.28 Another possible pattern (**a**) with a path through all nodes (**b**)

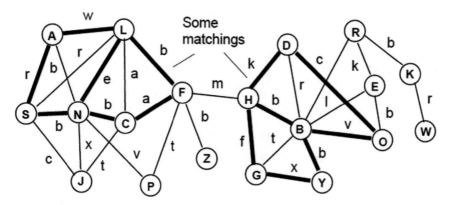

Fig. 5.29 Examples of matching pattern of Fig. 5.28 with the network

Fig. 5.30 Orders of accounting link names in the matching result

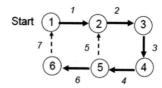

```
hop_nodes(all); frontal(Pattern) = NAME;
repeat_4(
    hop_links(all); notbelong(NAME, Pattern);
    Pattern && = NAME);
yes_hop(links(all), Pattern[2]);
hoplinks(all); notbelong(NAME, Pattern);
Pattern && = NAME;
hop(links(all), Pattern[1]); output(Pattern)
```

Some issued matchings, including those shown in Fig. 5.29, will be as follows:

```
(C,N,S,A,L,F), (G,H,D,O,B,Y), (S,A,L,C,J,N),
(A,L,C,N,J,S), …
```

We may also collect names of links in the matching cases found, putting them in a sequence with the order shown in Fig. 5.30 by numbers at arcs:

The modification of SGL scenario collecting link names too will be as follows:

Fig. 5.31 A pattern with
alternatives

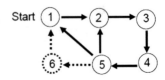

```
hop_nodes(all);
frontal(Pattern = NAME, Links);
repeat_4(hop_links(all); notbelong(NAME, Pattern);
        Pattern && = NAME; Links && = LINK);
hop(links(all), Pattern[2]);
Links && = LINK; hop(BACK);
hoplinks(all); notbelong(NAME, Pattern);
Pattern && = NAME; Links && = LINK;
hop(links(all), Pattern[1]); Links && = LINK;
output(Pattern, Links)
```

Examples of what will be output for a couple of matchings:

```
((A,L,F,C,N,S), (w,b,a,b,e,b,r)),
((G,H,D,O,B,Y), (f,k,c,v,b,b,x)), …
```

5.6.3   Search Patterns with Alternatives

Any search pattern, with any structure can be described in SGL and matched with
the whole network, often in parallel and distributed mode. Different strategies for
representing arbitrary graph patterns are possible (which may have alternatives, for
example).

In Fig. 5.31 a pattern is shown combining the two previous ones, with alternative
part leading from Figs. 5.26, 5.27 and 5.28 expressed by dotted lines.

Its SGL its coding will be as follows.

```
hop_nodes(all); frontal(Pattern) = NAME;
repeat_4(
   hop_links(all); notbelong(NAME, Pattern);
   Pattern && = NAME);
yes(hop(links(all), Pattern[2]);
branch(
   stay,
   (hop_links(all); notbelong(NAME, Pattern);
   Pattern && = NAME));
hop(links(all), Pattern[1]); output(Pattern)
```

The matching results will combine the two previous cases:

```
(A,L,F,C,N), (C,N,S,A,L), (D,B,Y,G,H), (G,H,D,O,B), …
(C,N,S,A,L,F), (G,H,D,O,B,Y), (S,A,L,C,J,N),
(A,L,C,N,J,S), …
```

There will be no problem to extend this scenario for collecting names of links too, as before.

We can also add more alternatives to the previous pattern, like the one shown in Fig. 5.32, allowing node 5 to directly follow node 3 by an alternative path, with the extended SGL scenario following.

```
hop_nodes(all); frontal(Pattern) = NAME;
repeat_2(
    hop_links(all); notbelong(NAME, Pattern);
    Pattern && = NAME);
branch(
    stay,
    (hop_links(all); notbelong(NAME, Pattern);
    Pattern && = NAME));
hop_links(all)); notbelong(NAME, Pattern);
Pattern && = NAME;
yes(hop(links(all), Pattern[2]));
branch(
    stay,
    (hop_links(all); notbelong(NAME, Pattern);
    Pattern && = NAME));
hop(links(all), Pattern[1]); output(Pattern)
```

With the latest extension, this united pattern, along with the cases shown in Figs. 5.27 and 5.29, will also be capable of discovering and collecting all structures of the type exhibited in Fig. 5.33 (only two such examples shown).

The united output under the pattern of Fig. 5.32 (combining cases shown in Figs. 5.27, 5.29 and 5.33) will be as follows:

(S,L,N,C), (B,R,E,O), (S,A,L,N), (S,J,C,N), (L,N,C,F),

(D,H,B,O), ...

(A,L,F,C,N), (C,N,S,A,L), (D,B,Y,G,H), (G,H,D,O,B), ...

(C,N,S,A,L,F), (G,H,D,O,B,Y), (S,A,L,C,J,N),

(A,L,C,N,J,S), ...

5.7 Physical Parameters of Distributed Social Systems

Social networks may have both virtual and physical dimensions, where their nodes, except names, may be associated with certain physical coordinates too, say, accessed

Fig. 5.32 A pattern with
more alternatives

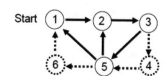

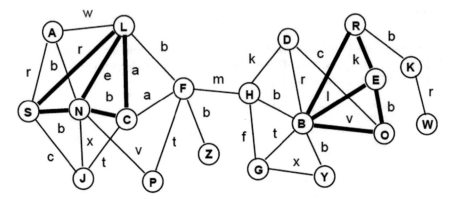

Fig. 5.33 Examples of additional matching by the pattern of Fig. 5.32

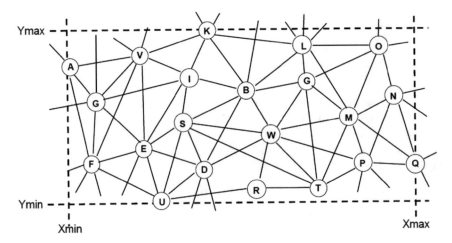

Fig. 5.34 Determining physical dimensions of a social network

by the environmental variables CONTENT in each node like X_Y value. Using the latter in different nodes, we may get general impression about the physical area covered by a concrete social network, as symbolically shown in Fig. 5.34 (In general, knowledge networks effectively processed in SGL may be borderless, covering the whole world, with thousands and millions of nodes and links which cannot be depicted together in one place like this figure.)

5.7.1 Determining Parameters of the Social Network Area

Finding lower and upper limits of two coordinates will be as follows, starting in some network node or from the network's outside:

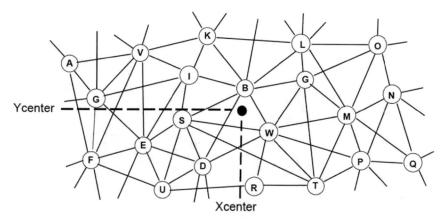

Fig. 5.35 Finding topographical center of the distributed network

```
nodal(Xmin, Xmax, Ymin, Ymax);
assignpeers((Xmin, Xmax),
                (min, max)(hop_nodes(all);CONTENT[1]));
assignpeers((Ymin, Ymax),
                (min, max)(hop_nodes(all);CONTENT[2]));
output(Xmin, Xmax, Ymin, Ymax)
```

And the general physical area covered by the social network, symbolically considered as the square one, will be having the following value:

```
output((Xmax - Xmin) * (Ymax - Ymin))
```

Any other physical features of the social network can be obtained too like, for example, the averaged topographical centre of the network (see Fig. 5.35), as follows:

```
nodal(Xcenter, Ycenter);
Xcenter = average(hop_nodes(all); CONTENT[1]);
Ycenter = average(hop_nodes(all); CONTENT[2]);
output(Xcenter, Ycenter)
```

We were using node coordinates of the virtual network presumably recorded in the environmental variable CONTENT associated with each node. But if network nodes from the beginning belong to the combination of virtual and physical worlds, i.e. having VPW origin (see Chaps. 2 and 3), their physical coordinates can be more naturally accessed by environmental variables WHERE in each combined node, rather than by CONTENT.

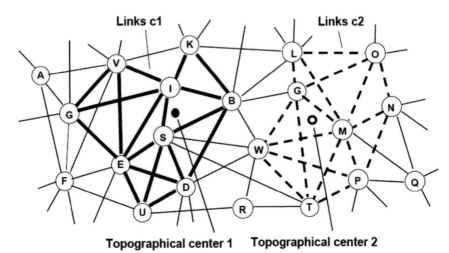

Fig. 5.36 Finding centers of different communities

5.7.2 Centers of Communities and Distances Between Them

Of practical interest may be finding topographical centers of different social communities with assessment of physical distances between them, say, for an intent of their integration or, on the other side, for preventing possible conflicts, as these groupings may be pursuing different, possibly, even hostile to each other, cultures, traditions, and principles. In Fig. 5.36, different communities are shown by different types of links between their members (like c1 and c2).

The following scenario outlines such communities by given types of links between their nodes, finds their topographical centers and evaluates and outputs physical distance between them (supposing that the extended rule average can work with two-dimensional coordinates at once).

```
nodal(Center1, Center2, Distance);
Center1 = average(hop_nodes(all);
                  yes(hop_links(c1)); WHERE);
Center2 = average(hop_nodes(all);
                  yes(hop_links(c2)); WHERE);
Distance = distance(Center1, Center2);
output(Distance)
```

The found Distance may, for, example, indicate some unwanted trends in a multicultural society if being below certain threshold, and may need special social and organizational measures to increase it (or, on the opposite, to decrease from another point of view).

We may also model some life and dynamics of the mentioned societies by allowing certain physical movement of their individual nodes in time (which may differ for different communities), with regular finding topographical centers of societies and measuring physical distance between them. We may also include issuing a warning if this distance becomes small and dangerous (or, for other applications, too large). This may be described by the following SGL scenario which has three independent distributed and repeating processes, with different individual delays after each iteration:

(a) continuous simulation of spatial dynamics in c1 community;
(b) continuous simulation of spatial dynamics in c2 community;
(c) continuous finding of the distance between topological centers of the communities with issuing a warning if becoming lower (or higher) than the threshold given.

```
nodal(Center1, Center2, Distance,
       Delay1 = ..., Delay2 = ..., Delay3 =...,
       Threshold = ..., Maxshift1 = dX1_dY1,
       Maxshift2 = dX2_dY2);
branch(
  (hop_nodes(all); yes(hop_links(c1));
  repeat(WHERE + random(Maxshift1);
         sleep(Delay1))),
  (hop_nodes(all); yes(hop_links(c2));
  repeat(WHERE + random(Maxshift2);
         sleep(Delay2))),
 repeat(
   Center1 = average(
     hop_nodes(all); yes(hop_links(c1)); WHERE);
   Center2 = average(
     hop_nodes(all); yes(hop_links(c2)); WHERE);
   Distance = distance(Center1, Center2);
   if(Distance <= Threshold,
       output('Danger' && Distance));
   sleep(Delay3)))
```

5.8   Conclusion

We have demonstrated some basic scenarios in SGL on social networks for analysing their structures. The offered solutions are ranging from evaluation of different types of nodes centrality, to discovering and dealing with network's weak and strong points with possibility of changing their topologies, to finding particular structures by given graph templates which may be of any complexity, to creation of arbitrary network topologies by self-spreading patterns, and finally, to assessing physical parameters of such networks and their parts with possible higher-level relationships between them, also simulation of their spatial dynamics.

The main difference of this work from any other existing activities on analysis and management of social networks is that all solutions exhibited in this chapter can be organized on arbitrary large social networks with may span any areas, the whole world including. These solutions, from the very beginning, are ideologically and conceptually formulated in most general parallel and fully distributed form, which allows us to effectively use for their implementation any scattered computational and communication resources available on the territories covered by the networks, and these resources can be engaged collectively and in parallel.

More on distributed processing of networks, social ones including, with the use of previous versions of the technology and language can be found in [12–20].

References

1. P. Sapaty, *Managing Distributed Dynamic Systems with Spatial Grasp Technology* (Springer, 2017)
2. P. Sapaty, *Ruling Distributed Dynamic Worlds* (Wiley, New York, 2005)
3. P. Sapaty, *Mobile Processing in Distributed and Open Environments* (Wiley, New York, 1999)
4. W. Fan, in *Graph Pattern Matching Revised for Social Network Analysis*, ICDT 2012, 26–30 March 2012 (Berlin, Germany). http://homepages.inf.ed.ac.uk/wenfei/papers/icdt12.pdf
5. C.T. Butts, Social network analysis: a methodological introduction. Asian J. Soc. Psychol. **11**, 13–41 (2008). http://courses.washington.edu/ir2010/readings/butts.pdf
6. C. Ni, C.R. Sugimoto, J. Jiang, *Degree, Closeness, and Betweenness: Application of Group Centrality Measurements to Explore Macro-disciplinary Evolution Diachronically*. https://www.elektrokomponenten.ch/media/files/09e4150bd20e58c913000000.pdf
7. S. Borgatti, *Centrality* (2005). http://www.analytictech.com/essex/lectures/centrality.pdf
8. D. Ortiz-Arroyo, Discovering sets of key players in social networks, in *Computational Social Network Analysis, Trends, Tools and Research Advances*, ed. by A. Abraham, A.-E. Hassanien, V. Snasel (Springer, 2010), pp. 27–47. http://pa.cm1911.com/Files/Subject/Computational%20social%20network%20analysis.pdf
9. *Node Centrality Measures* (1 Oct 2017). https://rampages.us/giny/2017/10/01/node-centrality-measures/
10. *What Are the Limitations of Graph Centrality Measures?* (26 June 2015). https://www.quora.com/What-are-the-limitations-of-graph-centrality-measures
11. L.C. Freeman, Centrality in social networks conceptual clarification. Soc. Netw. **1**(1978/79), 215–239 (Elsevier Sequoia S.A., Lausanne). http://leonidzhukov.net/hse/2014/socialnetworks/papers/freeman79-centrality.pdf

12. P.S Sapaty, WAVE-1: a new ideology of parallel processing on graphs and networks. Future Generations Comput. Syst. **4**, North-Holland (1998)
13. P.S. Sapaty, The WAVE-1: a new ideology and language of distributed processing on graphs and networks. Comput. Artif. Intell. **5** (1987)
14. P.S. Sapaty, A wave language for parallel processing of semantic networks. Comput. Artif. Intell. **5**(4) (1986)
15. P.S. Sapaty, *The Wave Approach to Distributed Processing of Graphs and Networks* (Proc. Int. Working Conf. Knowl. Vision Proc. Syst., Smolenice, 1986)
16. P.S. Sapaty, in *Active Information Field as a Model for Structural Solving of Tasks on Graphs and Networks*, Proceedings of the USSR Academy of Sciences. Technical Cybernetics, No. 5 (1984) (in Russian)
17. P.S. Sapaty, in *Solving Tasks on Semantic Networks and Graphs by Active Distributed Structures*, Proceedings of the 3rd International Conference Artificial Intelligence and Information-Control Systems of Robots, Smolenice, Elsevier Science Publishers B.V., North-Holland, 1984
18. P.S. Sapaty, in *On Efficient Structural Implementation of Operations on Semantic Networks*, Proceedings of the USSR Academy of Sciences. Technical Cybernetics, No. 5 (1983) (in Russian)
19. P.S. Sapaty, in *A Structural Approach to Solving Tasks of the Graph and Network Theory*, Proceedings of the 5th All-Union School on Parallel Programming and High-Performance Systems, Kiev, Naukova Dumka, 1982 (in Russian)
20. P. Sapaty, Distributed human terrain operations for solving national and international problems. Int. Relat. Diplomacy Vol. **2**(9) (2014)

Chapter 6
Real Networks Processing Examples

6.1 Introduction

The chapter shows some examples of dealing with realistic networks in SGL, with the use of network processing basics described in Chaps. 4 and 5. It starts with industrial networks with some analysis of Keiretsu networks dominant in Japan. Using vertical and horizontal features of such networks, a hypothetical distributed industrial-like network structure is offered on which exemplary operations in SGL are demonstrated.

A known and available on internet network example related to international business ecosystems is shown and analyzed too, having different types of companies-nodes and relations between them, like investment and partnership, with specific distributed operations on such networks demonstrated in SGL. Another considered example reflects known academic network in which persons with different scientific degrees and status communicate with each other, with exemplary operations on it, capable of discovering some interesting results like which kinds of persons communicate with other kinds and are requested most.

And finally, it stresses the difficulty even impossibility of analyzing and exhibiting very large industrial and social networks in a single point, which with numerous nodes and links often look together like "hairballs". With SGT, they can be effectively and holistically seen and analyzed in a completely distributed way, without any limitations on the number of nodes and links. Some typical operations on such networks are shown which may be useful even in full unawareness of their structure and semantics before contacting them.

The solutions shown in this chapter are based on the previous applications of SGT and SGL and their ancestor versions for solving different problems on graphs and networks, which are summarized in the preceding books [1–3].

© Springer Nature Switzerland AG 2019
P. S. Sapaty, *Holistic Analysis and Management of Distributed Social Systems*, Studies in Systems, Decision and Control 184, https://doi.org/10.1007/978-3-030-01830-6_6

6.2   Industrial Networks

6.2.1   Keiretsu Networks

We are starting here with Keiretsu networks [4–7]. Keiretsu is a Japanese word which, translated literally, means *headless combine*. It is the name given to a form of corporate structure in which a number of organizations link together.

A *keiretsu* (literally *system, series, grouping of enterprises, order of succession*) is a set of companies with interlocking business relationships and shareholdings. It is a type of informal business group. The keiretsu maintained dominance over the Japanese economy for the second half of the 20th century. The member companies own small portions of the shares in each other's companies, centered on a core bank. This system helps insulate each company from stock market fluctuations and takeover attempts, thus enabling long-term planning in projects. It is a key element of the manufacturing industry in Japan.

The primary aspect of the *horizontal Keiretsu* (also known as financial Keiretsu) is that it is set up around a Japanese bank, as in Fig. 6.1, where bank assists these companies with a range of financial services.

Typical of a Japanese horizontal Keiretsu is Mitsubishi where the Bank of Tokyo-Mitsubishi sits at the top of the Keiretsu. Also part of the core group is Mitsubishi Motors and Mitsubishi Trust and Banking followed by Meiji Mutual Life Insurance Company which provides insurance to all members of the keiretsu. Mitsubishi Shoji is the trading company for the Mitsubishi Keiretsu.

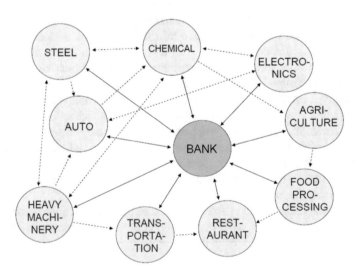

Fig. 6.1 Horizontal Keiretsu

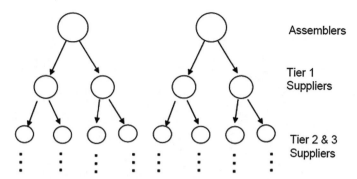

Fig. 6.2   Vertical Keiretsu

Vertical Keiretsu (also known as industrial Keiretsu) are used to link suppliers, manufactures, and distributors of the industry. In vertical Keiretsu, as in Fig. 6.2, one or more sub companies are created to benefit the parent company. Banks have less influence on vertical Keiretsu, and this vertical model is further divided into levels called *tiers*.

Vertical keiretsus are often the group of companies within the horizontal keiretsu such as Toyota. Toyota's success is dependent on suppliers and manufacturers for parts, employees for production, real estate for dealerships, steel, plastics and electronics suppliers for cars as well as wholesalers. All ancillary companies operate within the vertical keiretsu of Toyota but are members of the larger horizontal keiretsu, although much lower on the organizational chart.

6.2.2   Hypothetical Industrial Structure

In Fig. 6.3 a hypothetical distributed network, not directly corresponding to any existing network but rather expressing basic horizontal and vertical features (as well their integration) of Keiretsu-like industrial networks, is depicted, on which we will be trying to show some simple operations written in SGL. The core group members are represented as shaded `Ci` nodes with others being subsidiaries (just numbered) that may be hierarchically structured, with horizontal (`h`) and vertical (`v`) links between different nodes.

6.2.3   Some Basic Operations

Some scenarios expressing basic operations on such networks may be as follows.

- How many Core Group members (`Core`) are there in the network?

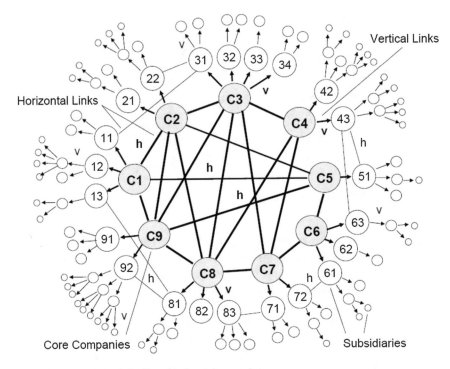

Fig. 6.3 A hypothetical distributed industrial network

```
output_count(hop_nodes(CONTENT == Core))
```

The answer in the location the scenario started will be: 9.

- List all Core Group members (Core).

```
output(hop_nodes(CONTENT == Core); NAME)
```

The answer in the starting position:

```
C1, C2, C3, C4, C5, C6, C7, C8, C9
```

- How many top level Subsidiaries (Subs) are there?

Starting from all Core nodes and then using vertical links v:

```
output_count(
        hop_nodes(CONTENT == Core); hop_link(v))
```

Or, by making the next hop to neighbors of Core nodes which are contented as Subs:

```
output_count(
        hop_nodes(CONTENT == Core);
        hop(links(any), nodes(CONTENT == Subs)))
```

Or, by hopping directly to all destination nodes if only top level subsidiaries are considered as Subs:

```
output_count(hop_nodes(CONTENT == Subs));
```

The answer in all three cases will be: 22.

- How many subsidiaries Core Group member C6 has on all its levels?

```
output_count(
        hop(C6); repeat(hop_links(+v); free(stay)))
```

Answer in the starting position: 11.

- Which Core Group member has most (and how many) subsidiaries (on all its levels)?

```
output_max(
    hop_nodes(CONTENT == Core);
    count_repeat(hop_links(+v); free(stay)) && NAME)
```

Answer in the starting position:

```
(17, C9)  or  (17, C4)
```

- Find all full groups of directly linked top subsidiaries by horizontal relations.

```
output(
        hop_nodes(CONTENT == Core); hop_links(+v);
        contain(
            IDENTITY = NAME; hopfirst(current);
            unit_repeat(
                if(NAME < IDENTITY, abort);
                free(NAME); hopfirst_links(h))))
```

Answers in the starting position:

```
(13, 81, 92), (11, 31,22), (43, 51, 63), (61, 72),
(13, 81, 92), (71, 83)
```

- Find all cliques among Core Group members linked by horizontal links h and having at least 4 nodes (see also Chap. 5 for similar solutions).

```
output(
  hop_nodes(CONTENT == Core);
  frontal(Clique) = NAME;
    contain(
        repeat(hop_links(h); notbelong(NAME, Clique);
                yes(and_parallel(
                        hop(links(any), nodes(Clique)))));
                if(PRREDECESSOR > NAME,
                    append(Clique, NAME), abort)));
  count(Clique) >= 4; Clique)
```

The answer in the position the scenario started will be:

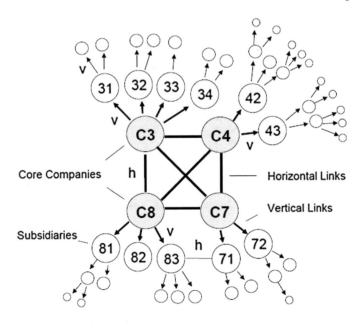

Fig. 6.4 Reduced industrial network

```
(C1, C2, C5, C9),  (C3, C4, C7, C8),
(C2, C3, C8, C9)
```

- Remove Core Group member reflected by node C5 together with other core nodes
  horizontally linked to it, and all subsidiaries, on all levels, associated with all these
  core nodes (this nonlocal operation, for example, may reflect a closure of related
  businesses, by a certain reason).

  This network reduction can be readily performed by the following parallel and
fully distributed SGL scenario which uses rule delete as a global context during
network navigation, with the resultant network shown in Fig. 6.4.

```
delete(hop(C5); hop_links(h), stay;
       repeat(hop_links(+v)))
```

6.3 Organizational Ecosystems

6.3.1 *Evolution of Business Ecosystems*

Organizations are now evolving into *organizational and business ecosystems* [8–12],
where organizational ecosystem is a system formed by the interaction of a community
of organizations and their environment. An ecosystem cuts across traditional industry

lines, and a company can create its own ecosystem. Some companies, for instance, may be active in several major industries, including consumer electronics, Internet services, mobile phones, personal computers, and entertainment. Their ecosystems also include multiple suppliers and numerous customers across many markets. Others are getting into entertainment business as well. Cable television companies are offering new forms of phone service, and telephone companies are getting into the television business. Today, successful companies develop relationships with numerous other organizations cutting across traditional business boundaries.

The essence of a business ecosystem is that networks between companies *need to be analysed from a higher* conceptual level rather than from the viewpoint of individual organisations. A business ecosystem's scope is the set of positive sum relationships between actors who work together around a core technology platform. Irrespective of an organisation's individual strength, all actors in a business ecosystem are connected and share the success or failure *of the network as a whole*.

6.3.2 International Business Network Example

International business network example is shown in Fig. 6.5 (this network picture, with some our edition, is based on 1999 data which is widely available on internet [9–11]). Not very fresh, however, but may nevertheless be useful for showing some basic network processing techniques in SGL for such types of organizations, as follows in this section.

Different types of business are shown in different colors in Fig. 6.5, these including:

1. System Manufacturing and Consumer Computing
2. Communication Services and Equipment (Wireless Including)
3. Systems Integration and Web Services
4. E-Commerce and Media
5. Data Networking
6. Biotechnology
7. Semiconductors
8. Software.

6.3.3 Possible Network Operations

Below we are offering some simple network analysis scenarios related to the network example shown in Fig. 6.5.

- How many organizations (nodes) are depicted in this network?

```
output_count(hop_nodes(all))
```

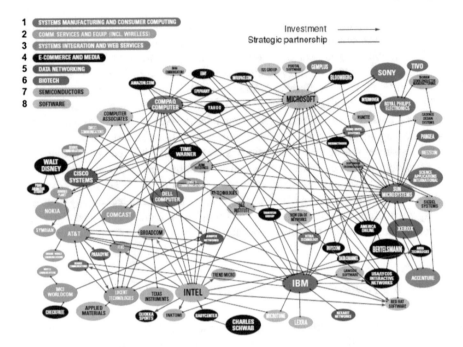

Fig. 6.5 Business ecosystem network example

Answer in the scenario application point (which can be any network node or a position outside the network): 79.

- Which organization has the highest number of links with other organizations, and the number of such links?

```
output_max(hop_nodes(all);
          count(hop_links(all)) && NAME)
```

Answer in the starting position will be:

```
22, SUN MICROSYSTEMS
```

- What is the minimal number of links from nodes to other nodes?

```
utput_min(hop_nodes(all); count(hop_links(all)))
```

Answer in the starting position: 1

- Determine minimal number of links emanating from nodes and then list all organizations with exactly the same number of connections.

```
frontal(Minlinks);
sequence(
   Minlinks = min(hop_nodes(all);
                  count(hop_links(all))),
```

```
output(hop_nodes(all);
        count(hop_links(all)) == Minlinks; NAME))
```

Answer in the starting position:

> ISS GROUP, PIVOTAL SOTYWARE, TIVO, INTERRWOVEN,
> EXODUS COMMUNICATIONS, PIXAR ANIMATION STUDIOUS,
> SYMBIAN, IRIDIUM WORLD COMMUNICATIONS, WHISTLE
> COMMUNICATIONS, CHECKFREE, APPLIED MATERIALS,
> QUOKKA SPORTS, TEXAS INSTRUMENTS, TREND MICRO,
> CHRLES SCHWAB, LEXRA, NEXABIT NETWORKS, SCIENCE
> APPLICATIONS INTERNATIONAL, BREEZECOM, PANGEA,
> EPIPHANY

- Which organizations have direct links with both SUN MICROSYSTEMS and MICROSOFT?

```
output(
    hop_nodes(all);
    yes_and(
        hop(links(any),
            nodes('SUN MICROSYSTEMS','MICROSOFT')));
    NAME)
```

Answer in the application point:

> VINCENTIE, WIND RIVER SYSTEMS, WEBMETHODS,
> TRINTECH GROUP, SAS INSTITUE, TRINTECH GROUP,
> NEW ERA OF NETWORKS, VITRIA TECHNOLOGY,
> AMERICA ONLINE

- Name all organizations with which INTEL has investment links.

```
output(hop_node(INTEL);
        hop_links(investment); NAME)
```

Answer in the application point:

> TREND MICRO, MICROSOFT, BABYCENTER, INKTOMI,
> QUOKKA SPORTS, NUANCE COMMUNICATIONS, ITXC,
> BROADCOM, ARIBA TECHNOLOGIES, RAD HAT SOFTWARE

- Find all paths between IBM and COMPAQ COMPUTER no longer than three (showing intermediate nodes only).

```
frontal(Path); hop_node(IBM);
 repeat_three(
     hop_links(all); notbelong(NAME, Path);
     if(NAME == 'COMPAQ COMPUTER', output(Path));
     Path &&= NAME)
```

All output of the found paths by this scenario will be in the final node COMPAQ COMPUTER as follows:

> (NUANCE COMMUNICATIONS, AT&T),
> (BROADCOM, AT&T), COMPUTER ASSOCIATES,

```
(AT&T, CISCO SYSTEMS), SAS INSTITUTE,
(NER ERA OF NETWORKS, MICROSOFT),
(TRINTECH GROUP, YAHOO), BROADCOM,
(TRINTECH GROUP, MICROSOFT),
(MICROTUNE, INTEL), AT&T,
(VITRIA TECHNOLGIES, MICROSOFT)
```

But with some elementary changes all this output can be organized in the scenario application point, with modified scenario looking like follows:

```
output(
       frontal(Path); hop_node(IBM);
       repeat_three(
          hop_links(all); notbelong(NAME, Path);
          if(NAME == 'COMPAQ COMPUTER', free_unit(Path));
          Path &&= NAME)
```

- Show all strategic partnership triangles.

```
hop_nodes(all); frontal(Triangle) = NAME;
repeat_two(
     hop_links(partnership); NAME < PREDECESSOR;
     Triangle &&= NAME);
hop(link(partnership, Trangle[1]);
output_unit(Triangle)
```

The following answers will be issued in nodes having maximum valued names among the three triangle nodes:

```
(CISCO SYSTEMS, COMPUTER ASSOCIATES, COMPAQ
COMPUTER),(COMPAQ COMPUTER, YAHOO, MICROSOFT),
(COMPAQ COMPUTER, TIME WARNER, MICROSOFT),
(COMPAQ COMPUTER, EBAY, MICROSOFT),
(AMAZON.COM, COPMPAQ COMPUTER, YAHOO),
(LUCENT TECHNOLOGIES, BROADCOM, COMPAQ COMPUTER),
(ITXC, AT&T, CISCO SYSTEMS),
(IT&T, CISCO SYSTEMWS, COMPAQ COMPUTER),
(ARM HOLDINGS, SUN MICROSYSTEMS, CADENCE DESIGN
SYSTEMS), (NUANCE COMMUNICATIONS, IBM, AT&T)
```

With slight changes, all such partnership triangles can be issued in the scenario starting position, as follows:

```
output(
     hop_nodes(all); frontal(Triangle) = NAME;
     repeat_two(
        hop_links(partnership); NAME < PREDECESSOR;
        Triangle &&= NAME);
     hop(link(partnership, Triangle[1]);
     unit(Triangle))
```

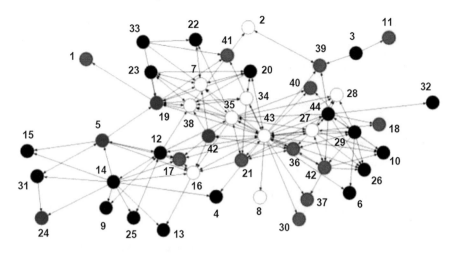

Fig. 6.6 Academic network example

6.4   Academic Networks and Operations

Social networks have also been applied to analysis to researchers' interactions in order to generate new metrics of scientific relations, which may add richness and differentiation to the scientific impact assessment [13–15].

An example of academic network describing communications between individual researchers with different scientific grades is shown in Fig. 6.6 (the network structure borrowed from [13] with our assignment of additional numerical node names necessary for the following analysis and discussions).

Nodes colouring in this network reflecting academic positions is as follows:

- black nodes = Ph.D.
- blue nodes = post-doc
- red nodes = full professor
- white nodes = assistant professor

Some exemplary scenarios for analysis of this network are presented below.

- List all full professors from the network.

```
output (
        hop_nodes (CONTENT == 'full professor'); NAME)
```

Answer in the application point:

```
        1, 17, 41, 21, 36, 30, 37, 42, 40, 39, 18, 11
```

- Find all linked groups of full professors who communicate with each other.

```
output (
    hop_nodes (CONTENT == 'full_professor');
```

```
contain(
 hopfirst(current);
 frontal(Start) = NAME;
 nodal(Group) = repeat(
     if(NAME > Start, abort); free(NAME);
     hopfirst(links(all),
              nodes(CONTENT == 'full professor'))));
count(Group) >= 2; unit(Group))
```

The answer will be issued in the scenario application point:

```
(21, 36, 42, 37)
```

Only one such group (of four) discovered, so looks like full professors communicate not much with each other, while prefer keeping distanced from same rank colleagues!

- Find the node which has most links with other nodes, and output the number of these links followed by node's scientific rank and its name.

```
output_max(
     hop_nodes(all);
     count_hop_links(all) && CONTENT && NAME)
```

Answer in the scenario application point:

```
23, assistant professor, 43
```

- Find which type of nodes has maximum communications with other node types, output this communication links number and the group's scientific rank.

```
output_max(
split('Phd', 'post-doc', 'full professors',
     'assistant professors');
frontal(Type) = VALUE;
count(
     hop_nodes(CONTENT == Type);
     hop(links(all),
         nodes(nonequal(CONTENT, Type)))) && Type)
```

Answer in the starting position:

```
42, 'assistant professor'
```

So from the previous two cases we may conclude that assistant professors communicate and also are requested (and definitely needed) mostly!

- Find all cliques with at least four nodes, uniting persons with possibly different scientific positions.

```
output(
  hop_nodes(all); frontal(Clique) = NAME;
  repeat(
    hop_links(all); not_belong(NAME, Clique);
    yes_and_parallel(hop(links(any), nodes(Clique)));
```

```
    if (BACK < NAME, Clique &&= NAME, done));
count (Clique) >= 4; unit (Clique))
```

Answer in the scenario staring position:

```
(26, 27, 29, 42, 44), (7, 35, 38, 42)
```

- Find the most isolated nodes having only a single or no link of communications with other nodes:

```
output (hop_nodes (all);
       count (hop_links (all)) <= 1; NAME)
```

Answer in the starting position:

```
1, 30, 11, 32, 8
```

- Find all Ph.D.'s who communicate with post-docs but not with full professors.

```
output (
   hop_nodes (CONTENT == PhD);
   no (hop_links (all); CONTENT == 'full professor');
   yes (hop_links (all); CONTENT == 'post-doc'); NAME)
```

Answer in the starting node or position:

```
23, 15, 31
```

6.5 Very Large Networks and Operations

Such networks [16–18] usually look like a "hairball" if many nodes and links are attempted to be exhibited on a single screen, being often completely unsuitable for any visual analysis, as in Fig. 6.7.

We will offer and consider here only very general operations on such networks, especially when knowing nothing in advance about their structures and semantics, which may be useful for their initial analysis and understanding. All output will be in a scenario starting position which can be a node inside the network or at its outside.

- How many nodes are there in the network?

```
output_count (hop_nodes (all))
```

- List all types of nodes:

```
output_unique (hop_nodes (all)); CONTENT)
```

- Output all types of nodes together with all node names belonging to each type:

```
nodal (Types, Nodes); frontal (Type);
Types = unique (hop_nodes (all)); CONTENT);
cycle (Type = withdraw (Types);
     Nodes = (hop_nodes (CONTENT == Type); NAME);
```

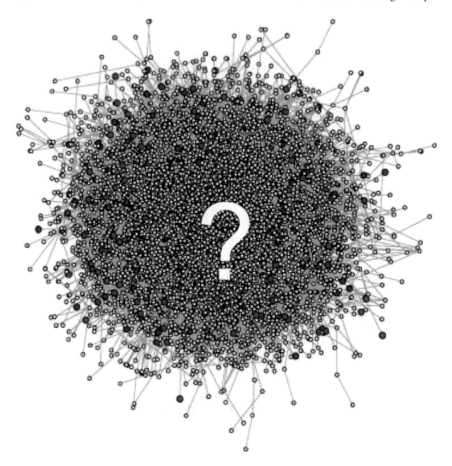

Fig. 6.7 Large networks, unsuitable for visual analysis

```
output_unit(Type, unit(Nodes))
```

- Create a new, additional, and much reduced network whose node names represent
  the discovered node types of the network of Fig. 6.7 (only one node for each node
  type), with a single link between any two of these new nodes if there exists at least
  one link with this name between nodes of the related types in the original network.

```
nodal(Types); frontal(Type1, Link, Type2)
Types = unique(hop_nodes(all)); CONTENT);
stay(create_nodes(Types); CONTENT = new);
hop_nodes_nonequal(CONTENT, new); Type1 = CONTENT;
hop(links(any),
    nodes_notbelong(CONTENT, (new, Type1)));
Link = LINK; Type2 = CONTENT; hop_node(Type1);
if(hop(link(Link), node(Type2)), quit,
    linkup(link(Link), node(Type2)))
```

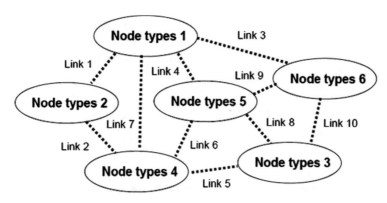

Fig. 6.8 The resultant network of interconnected node types

The resultant network summarizing node types of the network of Fig. 6.7 in the form of separate, united nodes, and links between the nodes inherited from the original network may look like the one shown in Fig. 6.8.

- List all types of links.

```
output_unique(hop_nodes(all)); hop_links(all);
        PREVIOUS > ADDRESS; LINK)
```

To guarantee that same links are accounted only once, we compare their two adjacent nodes by their addresses only, as large networks may have many same named nodes (this may, however, be redundant here as the scenario in any case will be collecting unique link names, omitting duplicates).

- Average number of links per node:

```
output_average(hop_nodes(all);
            count_hop_links(all))
```

- Finding a node with maximum links with outputting its name and a number of all links to other nodes. The output will be with only one such node, but there can be multiple nodes with this parameter. Below we will be finding all such nodes with maximum links.

```
output_max(hop_nodes(all);
            count(hop_links(all)) && NAME)
```

- Finding all nodes having the same maximum links with other nodes.

```
output(
  frontal(Maxlinks) =
          (hop_nodes(all); count_hop_links(all));
  hop_nodes(all);
  count_hop_links(all) == Maxlinks; NAME)
```

- List all isolated nodes.

```
output(hop_nodes(all);
        count_hop_links(all) == nil; NAME)
```

- Find the network diameter (as the number of links in the longest of shortest paths between any two nodes.

```
output_max(
      hop_nodes(all); IDENTITY = NAME;
      frontal(Far); nodal(Distance);
      stay_repeat(hop_links(all); increment(Far);
                  or(Distance == nil, Distance > Far);
                  Distance = Far);
   max(hop_nodes(all_other); Distance))
```

- Determine the whole physical area occupied by this network (if nodes have registered physical coordinates).

(a) Minimum, maximum on X coordinate:

```
output_apply((min, max)),
            (hop_nodes(all); WHERE[1]))
```

(b) Minimum, maximum on Y coordinate:

```
output_apply((min, max)),
            (hop_nodes(all); WHERE[2]))
```

(c) The whole physical area covered by the network:

```
nodal(Xmin, Xmax, Ymin, Ymax);
assignpeers((Xmin, Xmax),
      apply((min, max)), (hop_nodes(all); WHERE[1])));
assignpeers((Ymin, Ymax),
      apply((min, max)), (hop_nodes(all); WHERE[2])));
output('Area covered: ' &
                 (Xmax - Xmin) * (Ymax - Ymin))
```

Any more specific and complex requests, similar to many other examples in this book (especially with the use, if known, of any semantics associated with nodes and links which may be crucial for solving real problems) are possible under SGT.

6.6 Conclusion

We have shown examples of scenarios in SGL that can effectively operate on realistic networks which include known business, industrial and academic ones, also be very large and impossible to be investigated, processed and exhibited in any single point. Even when dealing with totally unknown networks, which may be arbitrarily distributed in unknown spaces too, SGT allows us to offer useful methods and solutions for their parallel access, investigation, abstraction, and even "seeing" under the spatial grasp mechanisms offered by the technology developed.

The SGL solutions are very compact as they semantically operate directly on the network surfaces, with omission of unnecessary details present in other models and languages,—by inheriting our brain's capabilities of unlimited scanning and holistic perception of complex network images. This is materialized in real distributed spaces by freely moving, self-replicating and self-matching intelligent scenario code enriched with different types of spatial variables and recursive track-based command and control.

More on distributed networks processing without limitations on the number of their nodes and structure, with the use of previous versions of the technology and language, can be found in [19–29].

References

1. P. Sapaty, *Managing Distributed Dynamic Systems With Spatial Grasp Technology* (Springer, Berlin, 2017)
2. P. Sapaty, *Ruling Distributed Dynamic Worlds* (Wiley, New York, 2005)
3. P. Sapaty, *Mobile Processing in Distributed and Open Environments* (Wiley, New York, 1999)
4. Economy of Japan. https://en.wikipedia.org/wiki/Economy_of_Japan
5. J. Grabowiecki, in *Keiretsu groups: Their Role in the Japanese Economy and a Reference Point (or a paradigm) for Other Countries*. V.R.F. Series, No. 413, Mar 2006. http://www.ide.go.jp/library/English/Publish/Download/Vrf/pdf/413.pdf

6. J.R. Lincoln, M. Shimotani, in *Whither the Keiretsu, Japan's Business Networks? How Were They Structured? What Did They Do? Why Are They Gone?* IRLE Working Paper No. 188-09 (2009). http://irle.berkeley.edu/workingpapers/188-09.pdf

7. A. Takeishi, Y. Noro, in *Keiretsu Divergence in the Japanese Automotive Industry: Why Have Some, But Not All, Gone?* CEAFJP Discussion Paper Series 17-04, CEAFJPDP, Aug 2017. http://ffj.ehess.fr/upload/Discussion/CEAFJPDP-17-04.pdf

8. J.F. Moore, Business ecosystems and the view from the firm. Antitrust Bull., Fall 2005. https://www.researchgate.net/publication/265217727_Business_ecosystems_and_the_view_of_the_firm

9. R.L. Daft. in *Organization Theory and Design*, 10th edn, South-Western, Cengage Learning (2010). https://www.kvimis.co.in/sites/kvimis.co.in/kumarfiles/ebook_attachments/Richard%20L.Daft%20Organization%20Theory%20and%20Design.pdf

10. R.L. Daft, in *Understanding the Theory and Design of Organizations*. South-Western, Cengage Learning (2010). https://www.amazon.com/Understanding-Theory-Design-Organizations-Richard/dp/1111826625

11. Organizational Ecosystems, Yagmurcevikgulblog, SOCI110 Module 5 - Interorganizational Relations (2017) http://yagmurcevikgul.blogspot.com/2017/10/organizational-ecosystems.html

12. Q.J. Zhao, Z.M. Wen, in *Integrative Networks of the Complex Social-Ecological Systems*. The 18th Biennial Conference of International Society for Ecological Modelling, Procedia Environ. Sci. **13**, 1383–1394 (Elsevier, New York, 2012). https://core.ac.uk/download/pdf/82297606.pdf

13. C.P. Hoffmann, C. Lutz, M. Meckel, in *Impact Factor 2.0: Applying Social Network Analysis to Scientific Impact Assessment*. 2014 47th Hawaii International Conference on System Science. https://ieeexplore.ieee.org/stamp/stamp.jsp?tp=&arnumber=6758799

14. P. Aventurier, in *Academic Social Networks: Challenges and Opportunities*, 7th UNICA Scholarly Communication Seminar Univ Sapienza Roma, 27–28 Nov 2014. http://www.unica-network.eu/sites/default/files/Academic_Social_Networks_Challenges_opportunities.pdf

15. M. Thelwall, K. Kousha, Academia.edu: Social Network or Academic Network? J. Am. Soc. for Inform. Sci. Technol. Apr 2014. https://www.researchgate.net/publication/256434230_Academiaedu_Social_Network_or_Academic_Network

16. A. Disney, Visualizing very large networks: an update, Cambridge Intelligence, 6 Dec 2016. https://cambridge-intelligence.com/visualizing-very-large-networks-an-update/

17. A. Clauset, M.E.J. Newman, C. Moore, Finding community structure in very large networks. Phys. Rev. **E70**, 066111, 6 Dec 2004. https://journals.aps.org/pre/abstract/10.1103/PhysRevE.70.066111

18. L. Lov´asz, Large networks and graph limits, Math. Subj. Classif. (2010). http://web.cs.elte.hu/~lovasz/bookxx/hombook-almost.final.pdf

19. P.S. Sapaty, in *A Structural Approach to Solving Tasks of the Graph and Network Theory*. Proc. 5th All-Union School on Parallel Programming and High-Performance Systems, Kiev, Naukova Dumka, 1982 (in Russian)

20. P.S. Sapaty, in *On Efficient Structural Implementation of Operations on Semantic Networks*. Proc. Technical Cybernetics, 5 (USSR Academy of Sciences Russian, 1983)

21. P.S. Sapaty, in *Active Information Field as a Model for Structural Solving of Tasks on Graphs and Networks*. Proc. Technical Cybernetics, 5 (USSR Academy of Sciences Russian, 1984)

22. P.S. Sapaty, A wave language for parallel processing of semantic networks. Comput. Artif. Intel. **5**(4) (1986)

23. P.S. Sapaty, in *The Wave Approach to Distributed Processing of Graphs and Networks*. Proc. Int. Working Conf. Knowledge and Vision Processing Systems, Smolenice, Nov 1986

24. P.S. Sapaty, The WAVE-1: a new ideology and language of distributed processing on graphs and networks. Comput. Artif. Intell. (5), 1987

25. P.S. Sapaty, WAVE-1: a new ideology of parallel processing on graphs and networks. Future Gener. Comput. Syst. **4**, North-Holland (1988)

26. P.M. Borst, M.J. Corbin, P.S. Sapaty, in *WAVE Processing of Networks and Distributed Simulation*. Proc. HPDC-3 Int. Conf., San Francisco, IEEE, pp. 61–69, Aug 1994

27. P.S. Sapaty, in *Solving Tasks on Semantic Networks and Graphs by Active Distributed Structures*. Proc. 3rd Int. Conf. Artificial Intelligence and Information-Control Systems of Robots, Smolenice, Elsevier, New York, B.V., North-Holland (1984)
28. P.S. Sapaty, Organization of advanced ISR systems by high-level networking technology. Math. Mach. Syst. (1) (2016)
29. P.S. Sapaty, in *Providing Over-Operability of Advanced ISR Systems by a High-Level Networking Technology*. SMI's Airborne ISR, Holiday Inn Kensington Forum, London, United Kingdom, 26–27 Oct 2015

Chapter 7
Robotized Societies

7.1 Introduction

Of particular interest and effectiveness may be approaches allowing for seamless embedment of massive robotics into human societies, with robots taking care of dangerous and critical situations while acting cooperatively with humans and among themselves under global goals and unified control. The chapter describes a novel approach for human-robot integration which is not pursuing traditional *interoperability* ideology and practice, but rather creating a higher *over-operability* layer in the form of supreme (i.e. standing above humans and robots) spatial intelligence. This layer allows us to express top semantics of what should be done in distributed spaces and main decisions to be taken regardless whether humans or robots engaged in this. Under this approach, it becomes easy to assemble any teams with any ratio between humans and robots, which can substitute each other at runtime without interrupting system missions while always preserving global goal orientation and mission capabilities.

The chapter briefs the philosophy and ideology of the over-operability concept which is technically supported by the Spatial Grasp Model, Language and Technology described in Chaps. 2 and 3 with combined human-robotic scenarios routinely organized and managed under this higher level. The considered scenarios include intelligent swarming observing some territory and written at different levels, from detailed to top semantic, where communicating group members can be humans, robots or both.

Managing emergency situations on roads with massive use of driverless, actually robotized, cars (manned ones can also be engaged), which behave collectively in a self-organized swarming mode, is considered too. The solutions written in SGL and explained in natural language include rational use of road space as regards vehicle's speeds, avoiding damaged vehicles, and car platoon management. The chapter also describes dynamic routing in road networks which can be useful for both manned

© Springer Nature Switzerland AG 2019
P. S. Sapaty, *Holistic Analysis and Management of Distributed Social Systems*, Studies in Systems, Decision and Control 184, https://doi.org/10.1007/978-3-030-01830-6_7

and unmanned cars, where vehicles, whether manned or unmanned, can chase each other, if needed, with the help of radar networks integrated with road infrastructures.

7.2 Human-Robotic Integration

7.2.1 Human-Robot Interoperability

Traditional models are treating both humans and robots, when considering their "interoperation", as separate entities, or "agents", exchanging messages, with individual human behaviors based on natural or Human Intelligence (HI), and those of robots on Artificial Intelligence (AI), as in Fig. 7.1. Interoperability is currently the dominant principle for joint operations in both civil and military areas [1–5], where human-robotic solutions are oriented on solving practical problems in real rather than abstract worlds, as shown in the figure.

Different kinds of interoperability are usually considered:

- *Syntactic interoperability* which means that two or more systems are just capable to communicate with each other.
- *Semantic interoperability* which supposes that beyond the ability to exchange information, different systems are capable of interpreting the exchanged information.
- *Cross-domain interoperability* which occurs when different kinds of entities (which may be multiple social, organizational, political, legal, etc.) can work together for a common purpose.

Due to the rapidly growing world dynamics and increasing complexity of joint operations, the existing interoperability principles, ideologically and primarily based on individual agents and their interactions, are becoming insufficient for providing

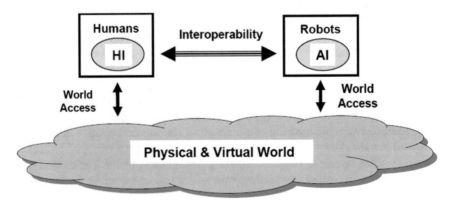

Fig. 7.1 Traditional human-robot cooperation model

the needed overall awareness, system integrity, and pursuit of global goals. We are often witnessing failures of the currently dominant interoperability-based approaches and solutions at national and international levels, where the whole of the problems is often underestimated and misunderstood, if not seen at all, after analyzing and implementing its parts first. This can also be true for advanced human-robotic collectives, which also need much higher level of integration than usual.

7.2.2 The Over-Operability Model

We have been developing a much higher formal model for dealing with distributed dynamic systems globally, and regardless of their origin, which may be human, robotic, social, economic, military, etc. This "over-operability" model [6–8], based on the Spatial Grasp approach and language described in Chaps. 2 and 3, also in previous books [9–11], conceptually consists on top level from two basic components: (a) integrated active physical, virtual and executive world, and (b) global, supreme, spatial intelligence staying above this world while overseeing and ruling it, with direct access to any world's points if needed, as in Fig. 7.2. This global intelligence formulates on top level what should be done in this world and which global decisions to be taken, presuming that self-organized world below can fulfil all this.

   In reality, the practical implementation of this over-operability model may need considering three levels after symbolically extracting part of the executive capabilities from the world with putting a new level of "intelligent doers" (which can include humans, robots, and any intelligent machinery) in between the previous two components, as shown in Fig. 7.3. This also results in adding to the top intelligent level certain capability of making high-level concretization and management, for proper interaction with this intermediate level, while always preserving its capability of directly influencing the remaining world beneath it without intermediaries.

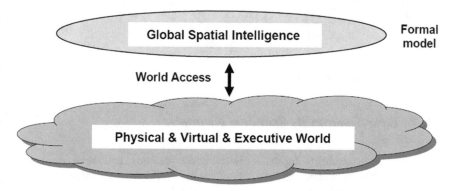

Fig. 7.2 Top level over-operability idea

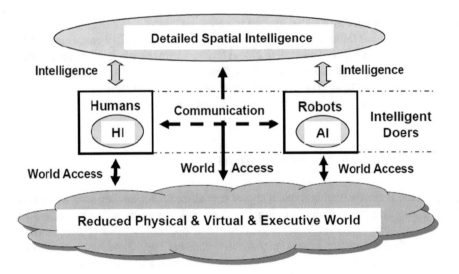

Fig. 7.3 Realistic over-operability organization

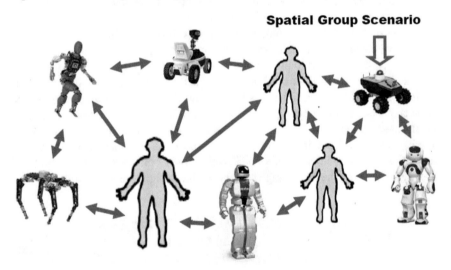

Fig. 7.4 Mixed human-robotic teams

7.2.3   Over-Operability Scenario Examples

Under the over-operability approach briefed in the previous section, we can easily organize any homogeneous or mixed collectives from different doers operating under the global, spatial, intelligence scenarios, as symbolically shown in Fig. 7.4.

These high-level scenarios are easily understandable to both humans and robots, and can be effectively executed by any homogeneous or heterogeneous groups, with some examples following. The highlighted (in bold) scenario parts can be executed by biological brain and sensors (say, in a dialog mode via a screen, keyboard, or voice, including what can be mounted on humans) if appear in SGL interpreters associated with humans, or entirely handled autonomously by robots otherwise, with overall group control fully automatic too.

7.2.3.1 Detailed Intelligent Swarming

Different stages of organization of collective movement, operation, and control of an intelligent swarm propagating in distributed physical space, which may be human, robotic or mixed, are shown below (see also [12, 13] for related scenarios).

- Randomized collective group movement, starting in any node, with minimal, Range distance allowed between units when moving.

```
hop(all);
nodal(Limits = (dx(0,8), dy(-2,5)), Range = 200,
      Shift);
repeat(Shift = random(Limits);
       if(empty(Shift, Range), WHERE += Shift);
       sleep(delay))
```

- Starting from any node, finding topologically central unit of the moving group (which may be changing in time due to the randomized movement of individual units and varying distances between units), and hopping into it.

```
frontal(Aver) = average(hop(all); WHERE);
min_destination(hop(all); distance(Aver, WHERE))
```

- Creating hierarchical infrastructure from the center found to other units using oriented links infra and parameter depth as a certain maximum allowed linking distance:

```
repeat_linkup(+infra, firstcome, depth))
```

- Using the created infrastructure, collect at its top unit (in a parallel bottom up procedure) and analyze all objects (symbolically: objects) discovered throughout the whole territory covered by the group, issuing resultant OK or alarm if danger.

```
frontal(Seen) =
   repeat(free_detect(objects), hop(+infra));
if(analyze(Seen), output(OK), output(alarm))
```

Integration of the above four cases within a single united scenario is trivial, as follows, allowing the whole group randomly move while keeping threshold distance between units, regularly redefining its changing center and hierarchical infrastructure stemming from it, also collection and analysis of objects seen.

```
frontal(Aver, Seen, Repetitions = …; Delay = …);
nodal(Limits = (dx(0,8), dy(-2,5)), Range = 200,
      Shift);
hop(all);
branch(
  repeat(Shift = random(Limits);
         if(empty(Shift, Range), WHERE += Shift);
    sleep(Delay)),
  repeat(
    Aver = average(hop(all); WHERE);
  min_destination(
      hop(all); distance(Aver, WHERE));
    stay(repeat_linkup(+infra, firstcome, depth)));
  loop_Repetitions(
    Seen =
      repeat(free_detect(objects); hop(+infra));
    if(analyze(Seen), output(OK), output(alarm);
    sleep(Delay))))
```

This scenario can start from any human or robotic unit, after which will be keeping full control over any movement and activity of the whole team, regardless of the number of units in it, which may be arbitrarily large. Any other collective scenarios can be generated too, often at runtime and on the fly, due to their transparency and compactness.

7.2.3.2 Top Level Semantic Scenario in SGL

At this highest level, it is possible to describe in SGL only what should be done in a distributed space and top decisions to make, like follows:

Evaluate damage after disaster in points with physical coordinates X1_Y1, X2_Y2, and X3_Y3, and report the obtained maximum damage value in the starting point.

The SGL expression for this scenario will be:

```
output_max_assess_damage(X1_Y1, X2_Y2, X3_Y3)
```

This semantic description is fully formal, and can be automatically implemented in physical space by any available manned, unmanned or mixed units. The solution by robotic units R1 and R2 and manned M1, accidentally scattered somewhere in the region (all having communicating SGL interpreters installed) is shown in Fig. 7.5.

The final result will be output in the node its was initially injected to, like robot R1 in Fig. 7.5, but this scenario can also start from any position outside this robotic-human environment, with final result to be output there too.

Other similar scenarios under the approach developed can be found in [10, 11, 14].

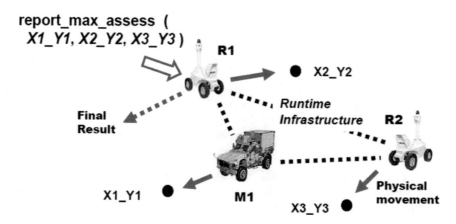

report_max_assess (
X1_Y1, X2_Y2, X3_Y3)

Fig. 7.5 Automatic solution of semantically defined problem

7.3 Societies with Driverless Cars

Autonomous vehicles [15–17] represent one of the most prominent technologies since the creation of automobile itself. World most influential companies are making billion-dollar investments in the driverless transport. Autonomous vehicles can fundamentally change transportation by reducing crashes, energy/fuel consumption, pollution and the costs of congestion. Human error is estimated to cause more than 90% of traffic accidents, a percentage that might be drastically reduced by the implementation of self-driving cars featuring smart systems that control most aspects of driving.

7.3.1 Car Autonomy Levels

An autonomous (driverless, self-driving, or robotic) car is a vehicle that is capable of sensing its environment and navigating without human input, and many such vehicles are being developed. Autonomous cars use a variety of techniques to detect their surroundings, such as radar, laser, GPS, odometry, and computer vision. Advanced control systems interpret sensory information to identify appropriate navigation paths and obstacles. The levels of vehicle's autonomy [18] are often classified as follows.

Level 0: The human driver controls everything like steering, brakes, throttle, power, etc.
Level 1: Most functions are still controlled by the driver, but steering or accelerating can be automated.
Level 2: Additionally to Level 1, cruise control and lane-centering can be automated. The driver can be disengaged from physically operating the vehicle, but still must always be ready to take control of the vehicle if needed.

Level 3: Drivers are still necessary, but can completely shift safety-critical functions to the vehicle under certain traffic or environmental conditions. The driver is still present but is not required to monitor the situation in the same way as for the previous levels.

Level 4: This means an essentially autonomous level where vehicles can perform all safety-critical driving functions and monitor roadway conditions for the entire trip. This, however, may not cover every driving scenario.

Level 5: This expects vehicle's performance being equal to that of a human driver, including extreme environments like, say, dirt roads.

Another challenging problem, not less complex that the development of autonomous cars itself, may be effective management of road networks with many driverless cars and considerable reduction of human involvement in the management process, on both local and global levels. This is an extremely difficult and so far unexplored task, which to a great extent will be determining the fate of this promising driverless field for the decades to come.

Using SGL, some basic emergent situations on roads with their solutions will be described and explained in detail, where vehicles can directly communicate with each other for finding suitable collective solutions using either short distance, direct V2V (vehicle-to-vehicle) communication channels, or for longer distances access the infrastructure levels by V2I (vehicle-to-infrastructure) types of communications (see Fig. 7.6).

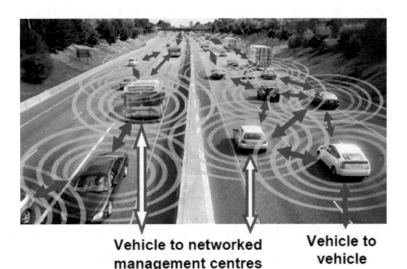

Vehicle to networked management centres

Vehicle to vehicle

Fig. 7.6 Communicating vehicles on the roads

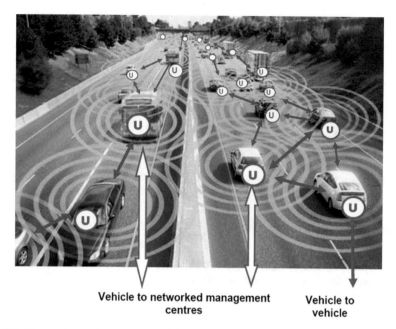

**Vehicle to networked management
centres**   **Vehicle to
vehicle**

Fig. 7.7   Cooperative road management with communicating SGL interpreters in vehicles

7.3.2   Examples of Collective Driverless Behaviour

Exemplary SGL scenarios of solving some very basic management problems on roads with autonomous cars, where such cars can communicate and cooperate with each other, will be shown below (see also [19]), assuming that every vehicle has individual SGL interpreter installed in it (as universal intelligent module U in Fig. 7.7, being an extension of Fig. 7.6). The scenarios can start from any car and dynamically involve as many vehicles as needed in its neighbourhood, and to any depth, covering the related regions with collective behaviours and dynamically creating spatial operational infrastructures. The latter can remain any period of time if continue serving the lasting solutions, otherwise automatically removed.

7.3.2.1   Narrowing the Gap Before Vehicle

Any vehicle seeing enough empty space before it (with certain given gap threshold value) and also getting information on the speed of the nearest vehicle ahead can update (like increase) its speed if its current speed and potential maximum speed can allow this. The vehicle can constantly make such checking in certain time intervals, with the corresponding solution shown in Fig. 7.8 (we will be using an improvised spatial sensing-communication-processing diagram, with d_i as current distances between neighbouring vehicles) and the SGL scenario code below.

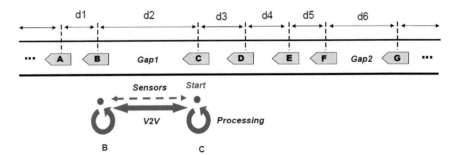

Fig. 7.8 Narrowing road gap before a vehicle

```
(Gap = …, Far, Speed_ahead);
sling(
  sleep(Delay);
  Far = distance(ahead, nearest) > Gap;
  Speed_ahead = (hop(ahead, nearest); SPEED);
  update(SPEED, (MAXSPEED, Speed_ahead, Far)))
```

Natural language comments on this compact and formal SGL scenario may be as follows.

- Define types and initial values of spatial variables used.
- Repeat the following regularly with certain time intervals.

  - Use sensors to measure nearest distance to the vehicles ahead.
  - Do the following if the gap ahead exceeds the threshold given.
  - Try to enter the nearest vehicle ahead via V2V, copy and bring back its current speed.
  - Update the current vehicle's speed taking into account its current speed, its possible maximum speed, current speed of the nearest vehicle ahead, and distance to this vehicle.

7.3.2.2 Narrowing Gap Before a Sequence of Vehicles

If a vehicle, like in the previous case, has a gap before it, and the directly following vehicle, as well as possibly all other following ones in the chain, have distances in between below threshold given, the speed changing operation can be done in all of them almost simultaneously, with the first one as the leader and all others trying to negotiate new speed in a direct contact with the previous vehicle. In a normal situation all vehicles behind the first one may refrain from changing their speed at this moment of time, but the first vehicle behind the gap may convince them this is proper time, and for a common benefit. This situation is shown in Fig. 7.9 and by SGL scenario that follows.

```
frontal(Gap = …, Speed_ahead, Far);
```

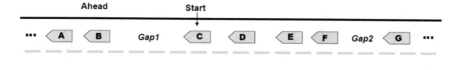

Fig. 7.9 Collective gap narrowing

```
sling(
  sleep(Delay);
  Far = distance(ahead, nearest) > Gap;
  Speed_ahead = (hop(ahead, nearest); SPEED);
  repeat(
    update(SPEED, (MAXSPEED, Speed_ahead, Far));
    Far = distance(behind, nearest) < Gap;
    Speed_ahead = SPEED;
    hop(behind, nearest)))
```

Informal natural language description of this scenario will be as follows.

- Define types and initial values of spatial variables used.
- Repeat the following regularly with certain time intervals.

  - Use sensors to measure nearest distance to vehicles ahead.
  - Do the following if the gap ahead exceeds threshold given, this vehicle declared to be the first and leading one.
  - Try to enter the first vehicle ahead via V2V, copy and bring back its current speed.
  - Do the following repeatedly until possible

    Update the current vehicle's speed taking into account its current speed, possible maximum speed, current speed of the vehicle ahead, and distance to the first vehicle ahead.
    Use sensors to measure distance to the nearest vehicles behind.
    If this distance is less than the given gap threshold, declare the current vehicle's speed as the speed of the vehicle ahead, enter the first vehicle behind via V2V, which becomes the current one now.

7.3.2.3   Lane Manoeuvring for Fastest Vehicle Between Gaps

A situation may occur that in a chain of vehicles with distances in between below threshold, like in the previous case, there may be vehicles with quite different maximum possible speed, and it could be reasonable to let the fastest one to make a solo lane manoeuvre and appear before the first vehicle behind the gap on the road. This situation is shown in Fig. 7.10 and by SGL scenario that follows, where, starting from the first vehicle, the whole chain of them is analyzed and the fastest vehicle found. The latter then making this manoeuvre individually, directly cooperating with the first vehicle it wants to be ahead of.

```
frontal(Gap = ..., Max, Before, First, Chosen);
sling(
   sleep(Delay);
   distance(ahead, nearest) > Gap; First = ADDRESS;
   repeat(
      if(MAXSPEED > Max,
         (Max = MAXSPEED; Chosen = ADDRESS));
      distance(behind, nearest) < Gap;
      hop(behind, nearest));
   hop(Chosen); lane_manoeuvre(ahead, First)))
```

A natural language explanation of this solution may be as follows.

- Define types and initial values of spatial variables used.
- Repeat the following regularly with certain time intervals.

  – Use sensors to measure the distance to nearest vehicle ahead and allow the following if the gap ahead exceeds threshold given.
  – Remember network address of the current vehicle, declaring it as the "first" one.
  – Do the following repeatedly until possible.
    If the current vehicle's maximum possible speed exceeds the already accumulated maximum speed among the considered vehicles, the latter changes to the former, and the current vehicle's network address is named as "chosen".
    Use sensors to measure distance to the nearest vehicle behind, and if it is less than the given gap threshold, enter the first vehicle behind via V2V, which becomes current now.

  – Directly enter the vehicle finally resulted as "chosen" and activate its lane manoeuvring, to appear ahead of the vehicle declared as "first", registered by its network address.

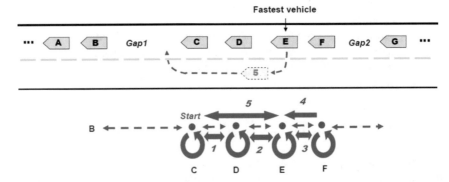

Fig. 7.10 Lane manoeuvring by the fastest vehicle

7.3.2.4 Collective Avoidance of Broken Car with Activity from It

A vehicle may accidentally stop, say, being damaged or somehow malfunctioning.
Assume also that this vehicle can still electronically communicate with other vehicles
on its initiative to inform them to avoid itself, and its distance to the nearest vehicle
ahead is big enough for a collective lane manoeuvring for the whole chain of vehicles
behind it. Such collective scenario launched by the initiative of the damaged vehicle
is depicted in Fig. 7.11 with the corresponding SGL code following.

```
frontal(Gap = …, Status);
sling(
  sleep(Delay);
  STATE == down; distance(ahead) > Gap;
  Status = head;
  repeat(
    distance(behind, first) < Gap;
    hop(behind, first);
    free(lane_manouvre(Status, BEFORE));
    Status = follower))
```

Natural language comments on this scenario may be as follows.

- Define types and initial values of spatial variables used.
- Repeat the following regularly and with certain time intervals.

  – If the vehicle's state is "down", and measured by sensors nearest distance to
    vehicles ahead exceeds the given threshold, allow the following.
  – Declare a possible next vehicle's status as "head" of a possible platoon-like
    chain of vehicles.
  – Do the following repeatedly until possible.

    Use sensors to measure distance to first vehicle behind, and if less than the given
    gap threshold, enter the first vehicle behind via V2V.

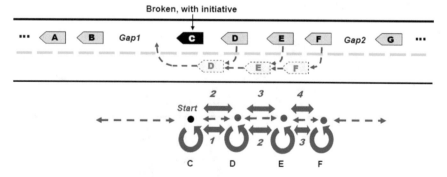

Fig. 7.11 Collective avoiding of the broken car by its initiative

Activate (in parallel with the rest of the scenario) lane manoeuvre procedure to avoid the damaged vehicle (with the "head" communicating with and orienting on the damaged one while others just interacting with and following the previous vehicles wherever they go and by which lane.

Setting the state of a possible next vehicle as "follower".

7.3.2.5 Collective Broken Car Avoidance with Activity Behind It

This scenario differs from the previous one by the fact that the damaged vehicle can still respond electronically but is unable to keep the whole initiative of activation and supervision of the following vehicles for a collective manoeuvre to become ahead of it. This initiative is instead delegated to the first vehicle behind the damaged one, which is becoming the leading vehicle for the whole operation, capable, however, of communication with the damaged vehicle to make the avoidance procedure as smooth as possible. This scenario is depicted in Fig. 7.12 with SGL code following.

```
frontal(Gap = …, Staus);
sling(
  sleep(Delay); distance(ahead, first) < Gap;
  hop(ahead, first); STATE == down;
  distance(ahead) > Gap; hop(BACK);
  Staus = head;
  repeat(
    free(lane_manouvre(Staus, BEFORE));
    distance(behind, first) < Gap;
    hop(behind, first);
    Staus = follower))
```

Natural language comments for this scenario are as follows.

- Define types and initial values of spatial variables used.
- Repeat the following and with certain time intervals.

  - If measured by sensors nearest distance to vehicles ahead is less than the given gap threshold, allow the following.

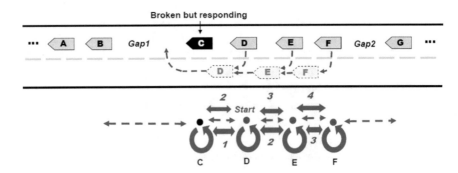

Fig. 7.12 Collective avoiding of the broken car by the initiative of the next vehicle

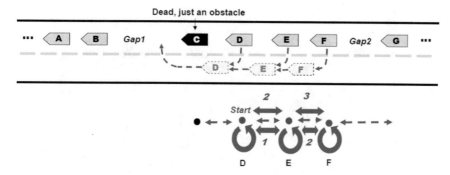

Fig. 7.13 Collective avoidance of a completely damaged vehicle

- Using V2V, hop to the nearest vehicle ahead, check its state, and if it is responding with "down" and measured distance to vehicles ahead of it is greater than the gap threshold given, allow the following.
- Hop back to the starting vehicle using V2V and name its position as "head".

- Do the following repeatedly until possible.
  Activate (in parallel with the rest of the scenario) the lane manoeuvre procedure to avoid the damaged vehicle (with the "head" vehicle communicating with and orienting on the damaged one while others interacting with and following the previous vehicles wherever they go and by which lane.
  Use sensors to measure distance to vehicles behind, and if less than the given gap threshold, enter the first vehicle behind via V2V and set its state as "follower".

7.3.2.6 Collective Avoidance of a Completely Broken Car

This case concludes that the damaged vehicle is completely "dead", actually becoming an obstacle to be fully ignored and avoided by all subsequent vehicles (i.e. not responding to the following vehicle trying to communicate with it electronically). The next to the damaged vehicle, instead of active dialogue with it, tries to determine its physical coordinates using sensors, which should be avoided when leading the chain of vehicles via another lane to appear ahead of the damaged one. This situation is shown in Fig. 7.13 and by SGL code below.

```
frontal(Gap = …); nodal(Place);
sling(
  sleep(Delay); distance(ahead) < Gap;
  or((hop(first_ahead); done),
     Place = measure(first_ahead));
  free(go_around(Place));
  repeat(
    distance(behind, first) < Gap;
    hop(behind, first);
    free(lane_manouvre(BEFORE))))
```

Fig. 7.14 Cars platooning on roads

Natural language comments for this scenario variation will be as follows.

- Define types and initial values of spatial variables used.
- Repeat the following and with certain time intervals.
  - If measured by sensors nearest distance to vehicles ahead is less than the given gap threshold, allow the following.
  - If using V2V to enter the nearest vehicle or obstacle ahead fails, use sensors to determine physical coordinates of the vehicle or obstacle ahead.
  - Organize parallel with the rest of the scenario lane manoeuvre for the current vehicle to go around the obstacle.
  - Do the following repeatedly until possible.

  Use sensors to measure distance to the first vehicle behind, and if less than the given gap threshold, enter this vehicle via V2V.
  Activate in it (in parallel with the rest of the scenario) the manoeuvre procedure to follow the previous vehicle wherever it goes and by which lane.

7.3.2.7   Collective Platoon Management

Platooning, a closely spaced multiple-vehicle chain on a highway (as in Fig. 7.14), has multiple benefits such as fuel saving, accident prevention, and so on. But it requires close cooperation among participating vehicles to maintain the platoon structure in case of different road situations.

(a) *Platoon management starting from its head*

The solution depicted in Fig. 7.15 and explained by the following SGL scenario, starting in the head vehicle, regularly accesses all vehicles in their chain while updat-

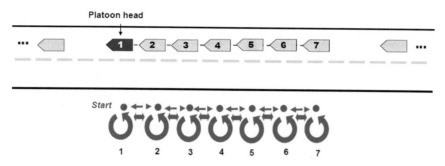

Fig. 7.15 Collective platoon management

ing their speed to keep the established standard distance between vehicles and orient the whole platoon on the speed of the head vehicle.

```
frontal(Distance = ..., Number = ..., Order = 1,
       Speed);
hop(Order);
sling(
  sleep(Delay); Speed = SPEED;
  repeat(
    Order += 1 <= Number;
    hop(behind, Order);
    update(SPEED,(Speed, Distance, BEFORE)))))
```

Explanation for this scenario will be as follows.

- Define types and initial values of spatial variables used.
- Start in the platoon's head vehicle.
- Repeat the following regularly and with certain time intervals.

    - Copy the head's current speed to be used for all other vehicles.
    - Repeat the following until possible.

    Increment vehicle's order to point to the next vehicle, not exceeding the total number of platoon vehicles.
    Enter the next vehicle by V2V with its ID to correspond to the current order.
    Update its speed taking into account the speed of the first vehicle and the needed and initially established distance between all vehicles in the platoon.

(b) *Collective management of a fragmented platoon*

Due to dynamic road conditions, traffic signals, road speed limits, and other factors like, for example, providing highest priority to emergency or police vehicles, a car platoon may suffer from fragmentation. Such a situation is depicted in Fig. 7.16, where between platoon vehicles 4 and 5 an emergency vehicle happened to appear on the same lane, which divided the platoon in two parts. For inclusion of such cases into platoon management, the previous scenario can be extended where in case of impossibility to contact next in line vehicle by V2V direct links, the current vehicle

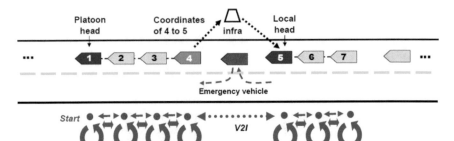

Fig. 7.16 Collective management of a fragmented platoon

uses more powerful V2I links with the road infrastructure. This is to find and contact the next vehicle which may happen to be at some distance or even far away, and transfer to it the current physical coordinates for the subsequent possibility of coming into the needed vicinity again, which will also be influencing all remaining platoon vehicles (i.e. vehicles 6 and 7). The corresponding SGL scenario code following.

```
frontal(Distance = …, Number = …, Order = 1,
        Speed, Position);
hop(Order);
sling(
  sleep(Delay); Speed = SPEED;
  repeat(
    Order += 1 <= Number;
    or(
      (hop(first_behind, Order);
       free(update(
          SPEED, (Speed, Distance, BEFORE))))),
      (Position = WHERE;
       hop(or(behind, infra), Order);
       free(update(Position)))))))
```

Natural language detailed explanation of this extended scenario is as follows.

- Define types and initial values of spatial variables used.
- Start in the platoon's head vehicle
- Repeat the following regularly and with certain time intervals.

  – Copy the head's current speed to be used in all other vehicles.
  – Repeat the following until possible.
    Increment vehicle's order to point to the next vehicle, not to exceed the total number of platoon vehicles.
    Two options possible:
    - *Option 1.* Enter next vehicle by V2V with its ID to correspond to the current order, and update its speed taking into account the speed of the first platoon vehicle and the needed and initially established distance between all vehicles in the platoon.

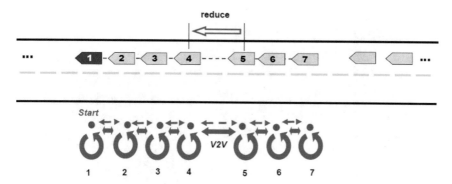

Fig. 7.17 Recovery of the platoon's structure

- *Option 2*. In case next vehicle is not accessible by V2V (and this may be when the platoon is broken into pieces being at some distance from each other), the physical coordinates of the current vehicle are copied and the next vehicle is tried to be contacted via more powerful V2I communications, and if successful, it is set to move to the coordinates of the previous vehicle.

(c) *Returning to normal platoon management after reducing gap between parts*

After vehicle 5 comes again into the reduced, normal distance after vehicle 4 by appearing as the next nearest vehicle to it, and the distance between them is covered by V2V communications, the whole platoon will be operating in one piece again, as shown in Fig. 7.17. This, however, will not guarantee the platoon of not being fragmented by certain forces and situations once more, so the previous scenario will be operating again for the platoon integration.

7.4 Autonomous Routing in Road Networks

Optimum routing is a complex problem in modern road networks (see Fig. 7.18), which will be even harder with the expected massive use of driverless cars.

7.4.1 Automatic Path Finding

A solution will be shown in SGL for finding shortest path between starting and final locations in distributed road networks of arbitrary complexity. The parallel spatial solution is based on directly navigating road infrastructure keeping the latest and runtime data on the roads availability, length, and current throughput, rather than

Fig. 7.18 Road networks

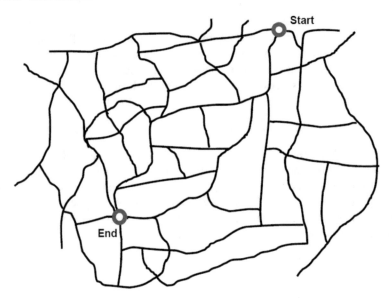

Fig. 7.19 Road network example

using predetermined maps. The parallel and fully distributed shortest path solution shown below is based on the general network solutions described in Chap. 4, also in [9–11]. A road network example with some Start and End points between which an optimum route is needed to be found is shown in Fig. 7.19.

An example of a supposed shortest path for this network within the allowed or restricted search area (as a road network may be very large and cannot be analysed in its entirety by the available time and resources) is shown in Fig. 7.20.

Possible parameters for defining the recommended search area are shown in Fig. 7.21, also mentioned below.

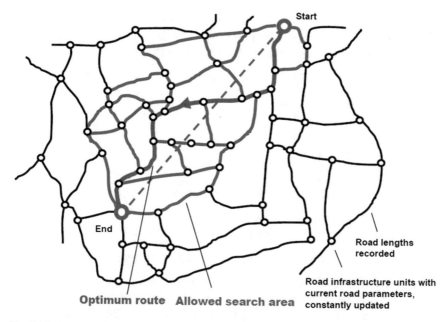

Fig. 7.20 An example of a possible shortest path within a restricted search area

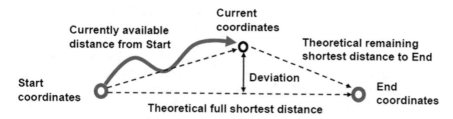

Fig. 7.21 Setting possible constraints on the shortest path solution

- Start and End nodes physical coordinates;
- Theoretical, or direct, distance between Start and End nodes;
- Already accumulated summary length of the considered path via the existing roads in the current node coordinates;
- Theoretical direct distance to the End node from the current node coordinates;
- Physical deviation of the current node coordinates from the direct line connecting Start and End nodes.

Using these parameters within the shortest path distributed algorithm of Chap. 4 (modified here), may allow us to restrict the shortest path procedure from the whole network to its reasonable area and speedup the solution while still providing its sufficient quality.

The following is a shortest path fully distributed solution (integrated with subsequent proceeding it by a vehicle physically) with taking into account the above mentioned restrictions narrowing and speeding the path search.

```
frontal(Far, Path, Start = startpoint;
        End = endpoint);
nodal(Distance, Passed);
hop_node(Start); Distance = 0; Path = NAME;
sequence(
 repeat(
   hop(road_infra, all); Far += LINK;
   acceptable(Far, WHERE, Start, End);
   or(Distance == nil, Far < Distance);
   Distance = Far; Path &&= NAME; Passed = Path;
   if(NAME == End, quit)),
 (Path = (hop_node(End); Passed);
  repeat(move_withdraw(Path, 1))))
```

Natural language comments to this scenario will be as follows.

- Define types and initial values of spatial variables used.
- Start in the current road infrastructure (RI) node using V2R connection.
- Proceed in a sequence the following two stages, each from the current RI node.

  – Stage 1. Repeat the following until possible by creating shortest path tree (SPT). Hop to all RI neighbouring nodes associated with neighbouring junctions. Bring into them the growing physical distance from the starting node incremented by the passed road length. If the evolving solution is within the permitted boundaries, and the brought distance is smaller than the one recorded in the node before, change the latter by the former and redefine the passed path in this node.
  – Stage 2. Organize physical car driving guided by the sequence of road junctions recorded in the path found and returned from the End node.

7.4.2 Automatic Chasing of Cars in a Road Network

Another task will be shown how advanced radars, which may be installed in numbers, say, in elevated positions and at main road junctions, as shown in Fig. 7.22, while communicating with each other, can trace suspicious cars wherever they go and inform authorities on their whereabouts (see also [20, 21]). Having regularly received the chased vehicles coordinates, police vehicles may try to reach these cars via the road network, optimally navigating the latter too.

Individual sensors/radars may have limited visibility range, but well organized distributed sensor networks empowered with SGT can provide continuous global vision of complexly moving objects through the space covered with their detailed study, also regularly informing certain control systems (like police) on their whereabouts, as shown in Fig. 7.23.

The SGL spatial tracking scenario guiding the operation of a distributed radar/sensor network can be as follows. On a global request, it catches the object (a car in our case) with the given identity and then follows wherever it goes with the help of individual mobile intelligence, regularly emitting its current physical coordinates to other systems. If the object cannot be seen from the current radar node any more (i.e. its visibility becomes lower than a given $threshold$), the scenario activates all neighboring radars and transfers full initiative and control to the radar seeing this object best, also informing if the object is eventually lost by the radar network.

```
frontal(Object) = identity;
nodal(Region) = limits;
max_fringe(hop_sensors(all, Region); seen(Object));
seen(Object) > threshold;
repeat(
    emit(Object && location(Object));
    loop(seen(Object) > threshold); sleep(delay));
    max_fringe(
        hop_sensors(all, neighbors); seen(Object));
    seen(Object) > threshold);
emit(Object, 'lost')
```

This scenario can be easily extended to the case where different mobile intelligence branches pursuing different moving objects simultaneously can cooperate with each other, possibly, under some global optimization processes (also in SGL). This may, for example, be to optimize the use of limited chasing/impact resources scattered throughout the region (like police or military vehicles) or recognize, analyze, and inform about group objects exhibiting a sort of collective behavior.

The use of regularly emitted data on the cars traced by radar network, say, by police vehicles trying to follow and reach suspicious cars (as in Fig. 7.24) may be presented in SGL as follows. (In the subsequent scenario, police cars are themselves

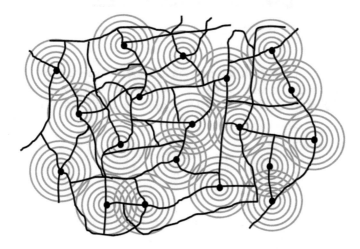

Fig. 7.22 Communicating radars implanted in a road network

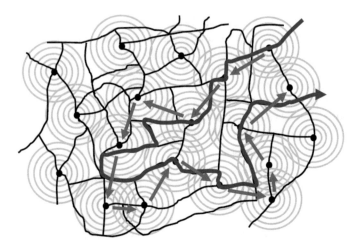

Fig. 7.23 Spatial tracing of a moving object in road network

Fig. 7.24 Tracing a car by police vehicles

trying to find shortest paths to the received locations of the suspected cars, which may change in time, i.e. with the need of runtime redefinition of shortest paths to the varying destinations, including abandoning of the obtained path-finding partial results).

```
hop(police_vehicle);
frontal(Start, End, Object, Far, Path);
nodal(Locus, Distance, Passed);
repeat(
  contain(
    get(Object, Locus);
    Start = WHERE; End = Locus; Distance = 0;
    stay_repeat(
      hop(road_infra, all); Far += LINK;
      acceptable(Start, End, Far, WHERE);
```

```
      or(Distance == nil, Far < Distance);
      Distance = Far; (Path, Passed) &&= NAME;
      if(NAME == End, quit));
   Path = (hop_node(End); Passed));
 repeat(
    get(Object, Locus);
    if(nonequal(End, Locus), abort);
    move_withdraw(Path, 1));
  abort)
```

7.5 Conclusion

While previous chapters were dealing with social systems formally represented as networks with graph-based operations on them, the current chapter brought more diversity and dynamics into social organisations by taking into account their rapidly evolving robotics dimension. Instead of traditional interoperability between humans and robots, where former used to task and control the latter, it discussed a much higher, or over-operability, model and layer staying above humans and robots and allowing for the expression of top semantics of operations and main decisions in distributed spaces regardless of who or what should be doing this.

Such approach allowed us to present high-level SGL scenarios which could be collectively executed by any mixture of human and robotics units, with any runtime swap between them. This vision also resulted in SGL solutions for exemplary, including emergency, situations on roads with massive involvement of driverless cars, the latter intelligently swarming under SGT. All shown and explained scenarios were extremely compact, transparent, and well understood to both humans and robots. Other human-robotic solutions under the over-operability approach developed can be found in [22–27].

References

1. G. Weichhart, M. Akerman, S.C. Akkaladevi, M. Plasch, A. Fast–Berglund, A. Pichler, in *Models for Interoperable Human Robot Collaboration*. Preprints of the 16th IFAC Symposium on Information Control Problems in Manufacturing Bergamo, Italy, 11–13 June 2018. https://d1keuthy5s86c8.cloudfront.net/static/ems/upload/files/eifig_0252_FI.pdf
2. W.A. Arokiasami, P. Vadakkepat, K.C. Tan, D. Srinivasan, Interoperable multi-agent framework for unmanned aerial/ground vehicles: towards robot autonomy. Complex Intell. Syst. **2**, 45–59 (2016). https://core.ac.uk/download/pdf/81067881.pdf
3. M. Aragão, P. Moreno, A. Bernardino, in *Middleware Interoperability for Robotics: A RoS—YARP Framework*. Frontiers in Robotics and AI, Published: 28 October 2016. https://pdfs.semanticscholar.org/b4ec/fd093fc40412be3cb48c91b678e0cc7d4f20.pdf
4. S.R. Ray, A.T. Jones, Manufacturing interoperability. J. Intell. Manufact. **17**(6), 535–540 (2003, January). https://www.researchgate.net/publication/221004522/download

5. J.G. Lofgren, in *NATO Capability Development Interoperability*. The Three Swords Magazine 30/2016. http://www.jwc.nato.int/images/stories/_news_items_/2016/LT_GEN_Lofgren_inte rview.pdf

6. P. Sapaty, in *Providing Over-operability of Advanced ISR Systems by a High Level Networking Technology*. SMI's Airborne ISR, 26th to 27th October 2015. Holiday Inn Kensington Forum, London, United Kingdom

7. P.S. Sapaty, Over-operability in distributed simulation and control. MSIAC's M&S J. Online **4**, 8 (2002)

8. P. Sapaty, Towards unified human-robotic societies. Austin J. Robot. Autom. **3**(1) (2017). http://austinpublishinggroup.com/robotics-automation/all-issues.php

9. P. Sapaty, *Mobile Processing in Distributed and Open Environments* (Wiley, New York, 1999)

10. P. Sapaty, *Ruling Distributed Dynamic Worlds* (Wiley Sons, New York, 2005)

11. P. Sapaty, *Managing Distributed Dynamic Systems with Spatial Grasp Technology* (Springer, Berlin, 2017)

12. P. Sapaty, M. Sugisaka, in *Optimized Space Search by Distributed Robotic Teams*. Proc. International Symposium on Artificial Life and Robotics (AROB 8th), Beppu, Japan, pp. 189–193, 24–26 January 2003

13. P. Sapaty, M. Sugisaka, in *Optimized Space Search by Distributed Robotic Teams*. Proc. World Symposium Unmanned Systems 2003, Baltimore Convention Center, USA, 15–17 July 2003

14. P. Sapaty, M. Sugisaka, in *Advanced Networking and Robotics for Societal Engagement and Support of Elders*. Proc. 16th International Symposium on Artificial Life and Robotics (AROB 16th '11), B-Con Plaza, Beppu, Oita, Japan (2011)

15. J.M. Anderson, N. Kalra, K.D. Stanley, P. Sorensen, C. Samaras, O.A. Oluwatola, Autonomous Vehicle Technology: A Guide for Policymakers. RAND Corporation, RR-443-2-RC, 214p (2016) https://www.rand.org/content/dam/rand/pubs/research_reports/RR400/RR443-2/RAN D_RR443-2.pdf

16. T. Litman, in *Autonomous Vehicle Implementation Predictions*. Implications for Transport Planning, Victoria Transport Policy Institute, 24 July 2018. http://www.vtpi.org/avip.pdf

17. L. Fridman, D.E. Brown, M. Glazer, W. Angell, S. Dodd, B. Jenik, J. Terwilliger, J. Kindelsberger, L. Ding, S. Seaman, H. Abraham, A. Mehler, A. Sipperley, A. Pettinato, B. Seppelt, L. Angell, B. Mehler, B. Reimer, MIT Autonomous Vehicle Technology Study: Large-Scale Deep Learning Based Analysis of Driver Behavior and Interaction with Automation, 19 Nov 2017. https://arxiv.org/pdf/1711.06976.pdf

18. Five Levels of Autonomous Driving. https://www.skoda-storyboard.com/en/innovation/five-l evels-autonomous-driving/

19. P. Sapaty, Distributed control technology for management of roads with autonomous cars. Int. J. Intell. Unmanned Syst. **5**(2/3) (2017). https://www.emeraldinsight.com/doi/full/10.1108/IJI US-05-2017-0006

20. F. Garcia, P. Cerri, A. Broggi, J.M. Armingol, A. Escalera, in *Vehicle Detection Based on Laser Radar*, ed by R. Moreno-D´ıaz et al. EUROCAST 2009, LNCS 5717 (Springer, Berlin, Heidelberg, 2009). https://pdfs.semanticscholar.org/a556/282836cce266ec24c6535d98f38d49 feb190.pdf

21. S. Adams, in *Vehicle Pursuits*. Legal Issues of Police Pursuits, Criminal Justice, Institute School of Law Enforcement Supervision, 3/7/2010. https://www.cji.edu/site/assets/files/1921/vehicle pursuits.pdf

22. P. Sapaty, Towards massively robotized systems under spatial grasp technology. J. Comput. Sci. Syst. Biol. **9**(1) (2016)

23. P. Sapaty, Unified transition to cooperative unmanned systems under spatial grasp paradigm. Int. J. Trans. Netw. Commun. (TNC), **2**(2) Apr 2014

24. P. Sapaty, in *From Manned to Smart Unmanned Systems: A Unified Transition*. SMi's Military Robotics, Holiday Inn Regents Park London, 21–22 May 2014

25. P. Sapaty, in *Human-Robotic Teaming: A Compromised Solution*. AUVSI's Unmanned Systems North America 2008, San Diego, USA, 10–12 June 2008
26. P. Sapaty, M. Sugisaka, in *Distributed Artificial Brain for Collectively Behaving Mobile Robots*. Proc. Symposium & Exhibition Unmanned Systems 2001, Baltimore, MD, 31 July–2 Aug 2001
27. P.S. Sapaty, in *Unified Transition to Robotized Armies with Spatial Grasp Technology*. International Summit Military Robotics, London, United Kingdom, 12–13 Nov 2012

Chapter 8
Gestalt-Based Distributed Vision Under SGT

8.1 Introduction

Gestalt psychology and theory [1–5] claim that human mind can directly grasp different images as a whole while interpreting their parts, which may be incomplete, in the context of this whole rather than vice versa. Gestalt principles are important to understand as they sit at the foundation of everything we do visually as designers. They describe how everyone visually perceives objects.

But using SGT, we can extend these gestalt features from traditional visual domain to "seeing" and understanding structures and situations distributed throughout large spaces and, moreover, do this remotely, which may be particularly important for advanced analysis and management of large distributed systems of different natures, social ones including.

The demonstration in SGL will include application of well known gestalt laws for dealing with exemplary spatial images placed, for convenience, on canvas of arbitrary large and distributed regular networks. These images will be analyzed and processed by SGL scenarios self-navigating and spatially matching them while simulating such gestalt laws as proximity, similarity, continuity, closure, common fate, symmetry, past experience, and good gestalt, as well as figure/ground concept and connectedness.

This chapter uses some ideas from our previous publications related to gestalt [6, 7], and all solutions presented are also based on the current and previous versions of SGT and its basic language along with their tested applications for distributed systems [8–10].

8.2 Extending Gestalt Laws to Distributed Systems

Gestalt is both philosophy and psychology term which means *unified whole* [1–5, 11]. It refers to visual perception theories developed by German psychologists at

© Springer Nature Switzerland AG 2019   185
P. S. Sapaty, *Holistic Analysis and Management of Distributed Social Systems*, Studies
in Systems, Decision and Control 184, https://doi.org/10.1007/978-3-030-01830-6_8

Fig. 8.1 Perceiving incomplete images as a whole

the beginning of the past century. These theories tried to understand and describe how people can organize visual elements into groups or unified wholes under certain conditions and principles, maintaining our meaningful perceptions in an apparently chaotic world.

The school of gestalt practiced a series of theoretical and methodological principles that attempted to redefine the existing approach to psychological research. This was in contrast to investigations developed at the beginning of the 20th century and based on traditional scientific methodology (still dominating at present!), which divided the object of study into a set of elements that could be analyzed separately with the objective of reducing the complexity of this object.

The word "Gestalt" is usually translated as a form, although it might be better understood as *organized structure*, as opposed to heap, aggregate, or simple summation. A few traditional perception examples are shown in Fig. 8.1 where we clearly understand the whole of these figures, their form, despite different parts missing.

Some author's personal experiments on assembling different shapes which could be perceived as a whole while resembling something or someone known before, regardless of the material used and their incompleteness, are shown in Fig. 8.2.

The wholes within the gestalt approach are considered as structured and organized using grouping laws, or *principles*. We will be trying to extend these very important gestalt laws in quite a new area for them—to the perception and management of large distributed systems, which may also include social networks discussed in this book. Will be doing this by representing, for convenience, distributed spaces as regular network structures shown in Fig. 8.3, with oriented horizontal and virtual links named, correspondingly, as h and v, and X_Y coordinates of nodes represented directly by their individual node names.

8.3   Expressing Law of Proximity

The **Law of Proximity** states that when individuals perceive an assortment of objects, they often first perceive objects that are close to each other as forming a group (as in Fig. 8.4). According to the law of proximity, things that are near each other seem to be grouped together.

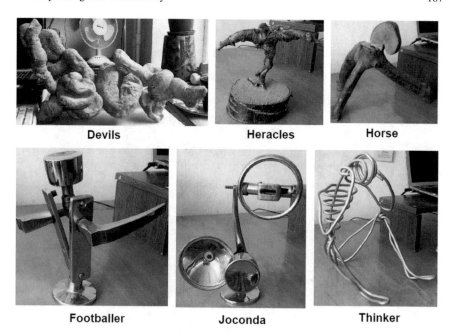

Fig. 8.2 Experiments on assembling of incomplete shapes

Let us distribute some round objects throughout the networked space as shown in Fig. 8.5 by giving X_Y coordinates of network nodes they should be placed into (these coordinates represented as node names), as follows.

```
hop_nodes(2_1, 2_10, 3_2, 3_7, 4_1, 4_2, 4_9, 5_1,
5_7, 5_12, 6_1, 7_9, 10_3, 10_4, 11_3, 11_4, 12_3,
12_4, 12_10, 12_11, 13_11, 13_12, 14_10 14_11);
CONTENT = circle
```

Let us try to find possible groups of these objects by synchronous breadth-first spanning tree coverage starting from a node and proceeding within the given threshold distance between nodes to be considered as belonging to a group, as follows.

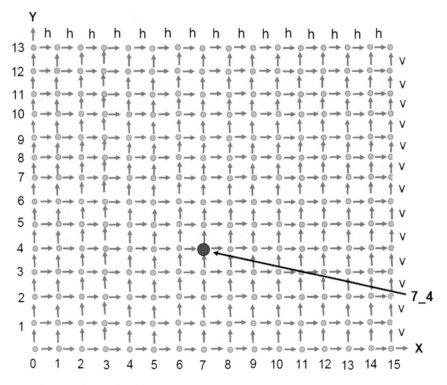

Fig. 8.3 Modelling distributed space by a regular network

Fig. 8.4 Proximity examples

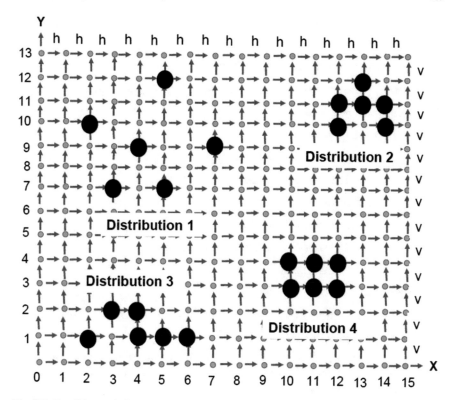

Fig. 8.5 Possible proximity examples

```
hop_nodes(all); nonempty(CONTENT); IDENTITY = NAME;
contain(
   frontal(Length, Step, Threshold = 5);
   nodal(Group, Distance, Proximity); Distance = 0;
   Group = NAME &&
      repeat_sync(
         (Step += 1) < Threshold;
         hop_links(all); Distance == nil;
         Distance = (Length += 1);
         if(nonempty(CONTENT),
            (if(NAME > IDENTITY, abort);
               free(NAME); Step = 0)));
   Proximity =
      sum(hop_nodes(Group); Distance)/count(Group);
   output(Proximity, Group))
```

Results for linking different nodes into possible groups of objects are shown in Fig. 8.6, with objects from which the groups started (chosen here as having maximum coordinate values, same as node names) are shown in red.

Calculated proximity results as averaged distances of nodes in their groups from the starting node, will be as follows (this is also compared with best possible results of group fixing if started from some other, more optimal nodes, which are closer to other group nodes).

$$\text{Proximity 1:} (3 + 5 + 4 + 6 + 6)/6 = 4$$
$$\text{best possible:} (3 + 4 + 3 + 3 + 3)/6 = 2$$

$$\text{Proximity 2:} (1 + 1 + 2 + 2 + 3)/6 = 1.5$$
$$\text{best possible:} (1 + 1 + 1 + 2 + 2)/6 = 1$$

$$\text{Proximity 3:} (1 + 2 + 3 + 4 + 4)/6 = 2.33$$
$$\text{best possible:} (1 + 1 + 2 + 2 + 2)/6 = 1.3$$

$$\text{Proximity 4:} (1 + 1 + 2 + 2 + 3)/6 = 1.5$$
$$\text{best possible:} (1 + 1 + 1 + 2 + 2)/6 = 1.17$$

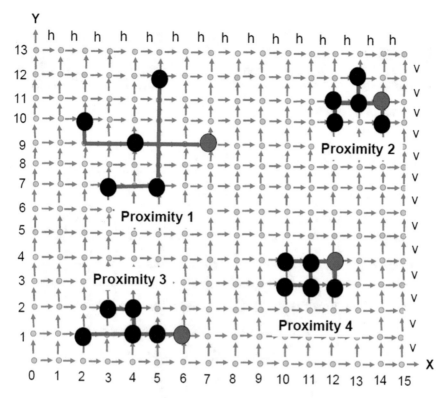

Fig. 8.6 Objects proximity fixed

The best proximity result will be for the competing groups named as Proximity 2 and Proximity 4. The best results did not influence the relative group ranks, with proximities 2 and 4 still remaining the best.

8.4 Expressing Law of Similarity

The **Law of Similarity** states that elements within an assortment of objects are perceptually grouped together if they are similar to each other. This similarity can occur in the form of shape, color, shading or other qualities, with some examples shown in Fig. 8.7.

By the following scenario we are placing different objects like triangle, circle, and square in different places on the regular grid as CONTENT of nodes, as shown in Fig. 8.8, and by the next scenario will be trying to find linked groups consisting of same objects.

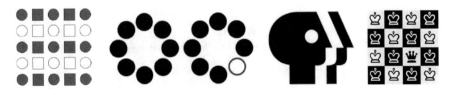

Fig. 8.7 Similarity examples

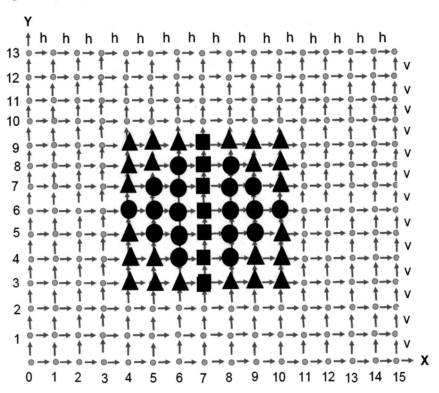

Fig. 8.8 Similarity to be found

```
parallel(
 (hop_nodes(
    4_3, 4_4, 4_5, 4_7, 4_8, 4_9, 5_3, 5_4,
    5_8, 5_9, 6_3, 6_9, 8_3, 8_9, 9_3, 9_4, 9_8, 9_9,
    10_3, 10_4, 10_5, 10_7, 10_8, 10_9);
  CONTENT = triangle),
 (hop_nodes(
    4_6, 5_5, 5_6, 5_7, 6_4, 6_5, 6_6, 6_7,
    6_8, 8_4, 8_5, 8_6, 8_7, 8_8, 9_5, 9_6, 9_7,
    10_6);
  CONTENT = circle),
 (hop_nodes(7_3, 7_4, 7_5, 7_6, 7_7, 7_8, 7_9);
  CONTENT = square))
```

Similarity groups finding scenario in SGL will be as follows (there can by be more than one tightly linked group having same objects):

```
hop_nodes(all); nonempty(CONTENT);
contain(
    IDENTITY = NAME; frontal(Object) = CONTENT;
    nodal(Group); hopfirst(current);
    Group = NAME &&
       repeat(hopfirst_links(all); CONTENT == Object;
                if(NAME < IDENTITY, abort); free(NAME));
    output(Object, unit(Group)))
```

The discovered groups consisting of different objects are exhibited by different colours in Fig. 8.9, with existence of different groups with same colours.

The output naming all these groups, performed in nodes starting them (having maximum value as name-coordinate in the group, to exclude duplicates from forming and registration of groups) will be as follows:

Group 1: `triangle, (6_9, 5_9, 5_8, 4_9, 4_8, 4_7)`
Group 2: `triangle, (10_9, 10_8, 10_7, 9_9, 9_8, 8_9)`
Group 3: `triangle, (6_3, 5_4, 5_3, 4_5, 4_4, 4_3)`
Group 4: `triangle, (10_5, 10_4, 10_3, 9_4, 9_3, 8_3)`
Group 5: `circle, (6_8, 6_7, 6_6, 6_5, 6_4, 5_7, 5_6, 5_5, 4_6)`
Group 6: `circle, (10_6, 9_7, 9_6, 9_5, 8_8, 8_7, 8_6, 8_5, 8_4)`

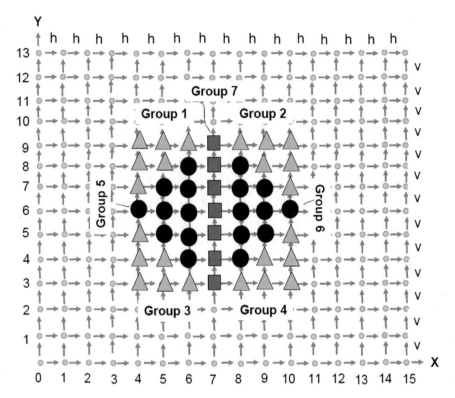

Fig. 8.9 Similarity finding results

Group 7:   square, (7_9, 7_8, 7_7, 7_6, 7_5, 7_4, 7_3)

8.5   Expressing Law of Continuity

The **Law of Continuity** states that elements of objects tend to be grouped together and integrated into perceptual wholes if they are aligned within an object. In cases where there is an intersection between objects, individuals tend to perceive the two objects as two single uninterrupted entities. It's instinct to follow a river, a path or a fence line. Once you look or move in a particular direction, you continue to look or move in that direction until you see something significant or you determine there's nothing significant to see. See Fig. 8.10 for usual continuity examples.

We will consider here only very simple examples, starting from straight lines which may intersect, as in Fig. 8.11. Initial placing objects into different positions of our regular network can be similar to how we did this before.

The subsequent scenario first finds the ends of the lines and then, determining their direction, follows them while ignoring possible intersection with other lines.

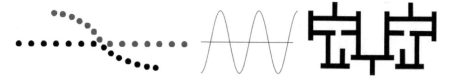

Fig. 8.10 Continuity examples

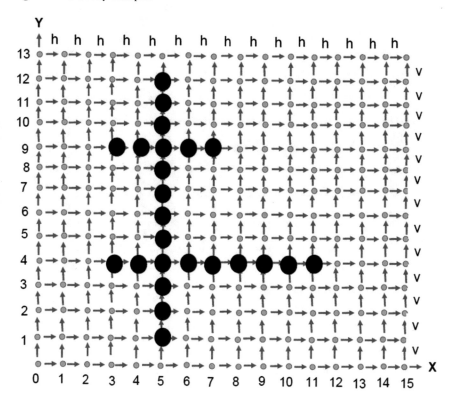

Fig. 8.11 Simple continuity cases

```
hop_nodes(all); nonempty(CONTENT);
count(hop_links(all); nonempty(CONTENT)) == 1;
frontal(Link) = all; nodal(Group, Threshold = 3);
Group = NAME &&
    repeat(hopforth_link(Link); nonempty(CONTENT);
            Link = LINK; free(NAME));
count(Group) > Threshold; output('Line:', Group)
```

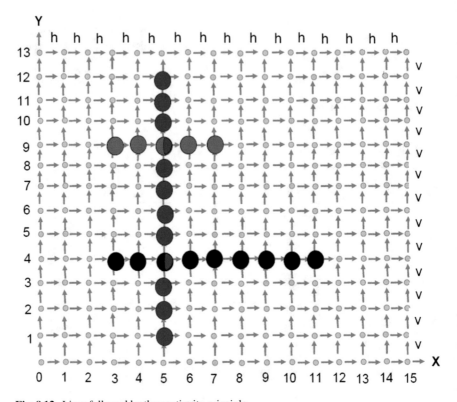

Fig. 8.12 Lines followed by the continuity principle

The lines followed by this scenario are shown as in Fig. 8.12 (they are marked in different colours).

The results output in nodes the lines started will be as follows:

- Starting from 5_1:

        Line: (5_1, 5_2, 5_3, 5_4, 5_5, 5_6, 5_7, 5_8, 5_9,
               5_10, 5_11, 5_12)

- Same line, but starting from its other end 5_12, and moving in the opposite direction:

        Line: (5_12, 5_11, 5_10, 5_9, 5_8, 5_7, 5_6, 5_5,
               5_4, 5_3, 5_2, 5_1)

- Starting from 3_9:

        Line: (3_9, 4_9, 5_9, 6_9, 7_9)

- Same line, but starting from its other end 7_9 and moving in the opposite direction:

        Line: (7_9, 6_9, 5_9, 4_9, 3_9)

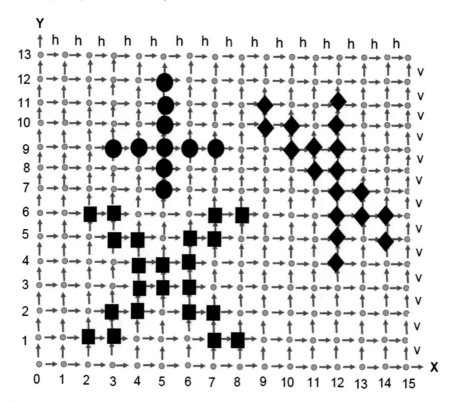

Fig. 8.13 Some more general continuity cases

- Starting from 3_4:

    Line: (3_4, 4_4, 5_4, 6_4, 7_4, 8_4, 9_4, 10_4,
           11_4)

- Same line but starting from its other end 11_4 and moving in the opposite direction:

    Line: (11_4, 10_4, 9_4, 8_4, 7_4, 6_4, 5_4, 4_4,
           3_4)

Some more general cases, containing diagonal lines mapped on the chosen rect-angular grid are shown in Fig. 8.13, where placing different objects in the network nodes will be similar to the previous examples.

The following scenario will be serving these extended examples on Fig. 8.13.

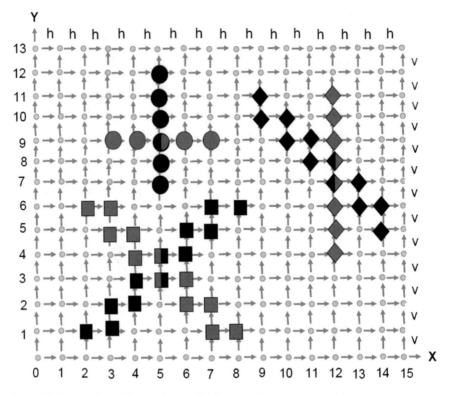

Fig. 8.14 Lines followed in extended continuity examples

```
hop_node(all); nonempty(CONTENT);
count(hop_link(all); nonempty(CONTENT)) == 1;
frontal(Link1, Link2) = all;
nodal(Group, Threshold = 6);
Group = NAME &&
 repeat(hopforth_link(Link1); nonempty(CONTENT);
        Link1 = DIRECTION & LINK; free(NAME);
        hopforth_link(Link2); nonempty(CONTENT));
        Link2 = DIRECTION & LINK; free(NAME));
count(Group) > Threshold; output('Line: ', Group)
```

The lines pursued by this scenario are highlighted by different colours in Fig. 8.14. The output of the collected lines will be in nodes the lines started, as follows:

- Starting from 5_12:

      Line: (5_12, 5_11, 5_10, 5_9, 5_8, 5_7)

- Same line but starting from 5_7 and moving in opposite direction:

```
Line: (5_7, 5_8, 5_9, 5_10, 5_11, 5_12)
```

- Starting from 3_9:

```
Line: (3_9, 4_9, 5_9, 6_9, 7_9)
```

- Same line but starting from 7_9 and moving in opposite direction:

```
Line: (7_9, 6_9, 5_9, 4_9, 3_9)
```

- Starting from 12_11:

```
Line: (12_11, 12_10, 12_9, 12_8, 12_7, 12_6, 12_5,
       12_4)
```

- Same line but starting from 12_4 and moving in opposite direction:

```
Line: (12_4, 12_5, 12_6, 12_7, 12_8, 12_9, 12_10,
       12_11)
```

- Starting from 9_11:

```
Line: (9_11, 9_10, 10_10, 10_9, 11_9, 11_8, 12_8,
       12_7, 13_7, 13_6, 14_6, 14_5)
```

- Same line but starting from 14_5 and moving in opposite direction:

```
Line: (14_5, 14_6, 13_6, 13_7, 12_7, 12_8, 11_8,
       11_9, 10_9, 10_10, 9_10, 9_11)
```

- Starting from 2_6:

```
Line: (2_6, 3_6, 3_5, 4_5, 4_4, 5_4, 5_3, 6_3, 6_2,
       7_2, 7_1, 8_1)
```

- Same line but starting from 8_1 and moving in opposite direction:

```
Line: (8_1, 7_1, 7_2, 6_2, 6_3, 5_3, 5_4, 4_4, 4_5,
       3_5, 3_6, 2_6)
```

- Starting from 2_1:

```
Line: (2_1, 3_1, 3_2, 4_2, 4_3, 5_3, 5_4, 6_4, 6_5,
       7_5, 7_6, 8_6)
```

- Same line but starting from 8_6 and moving in opposite direction:

```
Line: (8_6, 7_6, 7_5, 6_5, 6_4, 5_4, 5_3, 4_3, 4_2,
       3_2, 3_1, 2_1)
```

- Starting from 2_6:

```
Line: (2_6, 3_6, 3_5, 4_5, 4_4, 5_4, 5_3, 6_3, 6_2,
       7_2, 7_1, 8_1)
```

- Same line but starting from 8_1 and moving in opposite direction:

```
Line: (8_1, 7_1, 7_2, 6_2, 6_3, 5_3, 5_4, 4_4, 4_5,
       3_5, 3_6, 2_6)
```

Fig. 8.15 Closure examples

8.6 Expressing Law of Closure

The **Law of Closure** states that individuals perceive objects such as shapes, letters, pictures, etc., as being whole even when they are not complete. Specifically, when parts of a whole picture are missing, our perception fills in the visual gap, with related examples shown in Fig. 8.15.

In Fig. 8.16, a square shape with missing parts is shown which is mapped on our regular network canvas, and we will show how to fill in the gaps. Initial placing objects to network nodes as in Fig. 8.16 will be similar to what we did before. Filling the gaps on sides of the square shape and in corners can be done differently by the following scenario.

```
frontal(Figure == circle, Link);
nodal(Mark, Threshold = maxdistance);
sequence(
   (hop_nodes(all); CONTENT == Figure;
     Link = (hop_link(all); CONTENT == Figure;
                reverse(DIRECTION) & LINK);
      count(Link) == 1; Mark = Link;
      repeat_Threshold(
          hop_link(Link);
          or_sequence(
              (empty(CONTENT); CONTENT = 1),
              (CONTENT == 1; blind(CONTENT = Figure))))),
   (hop_nodes(all); nonempty(Mark); Link = Mark;
      repeat(hop_link(Link); CONTENT == 1;
          CONTENT = Figure)))
```

Two stages of this scenario (each being parallel), operate in a sequence, where after the first stage it may happen that not all missing positions close to corners are filled in but all corners have been inserted. At the next stage all remaining positions at the sides which are close to corners will be filled. Some stages of this scenario are depicted in Figs. 8.17 and 8.18.

We can organize final check of the completeness of recovered shape (the latter shown in Fig. 8.19) by the following scenario:

```
hop_random_nodes(CONTENT == circle);
frontal(Shape) = NAME;
repeat(
```

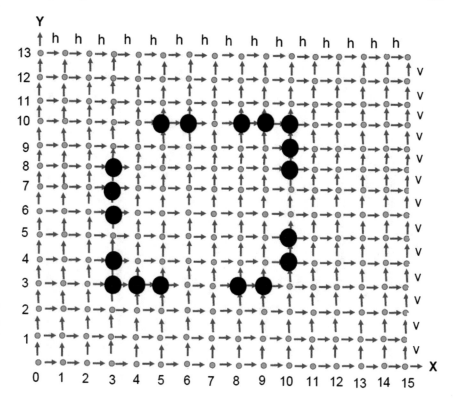

Fig. 8.16 Closure to be found

```
hopforth_links(any); CONTENT == circle;
if(NAME == Shape[1],
   blind_output(count(Shape), Shape));
Shape && = NAME)
```

Printed result in the randomly chosen node (being both starting and final one) will include calculation of how many nodes with circles constitute the full shape, also listing names (as X_Y addresses) of all nodes forming the shape:

```
28, (6_10, 7_10, 8_10, 9_10, 10_10, 10_9, 10_8,
10_7, 10_6, 10_5, 10_4, 10_3, 9_3, 8_3, 7_3, 6_3,
5_3, 4_3, 3_3, 3_4, 3_5, 3_6, 3_7, 3_8, 3_9, 3_10,
4_10, 5_10)
```

8.7 Expressing Law of Common Fate

The **Law of Common Fate** states that objects are perceived as moving together. Experiments using the visual sensory modality found that movement of elements of an object produce paths that individuals perceive that the objects are on.

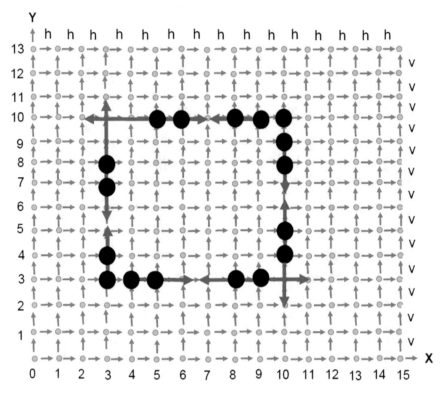

Fig. 8.17 Filling gaps by distributed processes with direction shown by arrows

When visual elements are seen moving in the same direction at the same rate, perception associates the movement as part of the same stimulus. For example, birds may be distinguished from their background as a single flock because they are moving in the same direction and at the same velocity, even when each bird is seen—from a distance—as little more than a dot. The moving 'dots' appear to be part of a unified whole. Similarly, two flocks of birds can cross each other in a viewer's visual field, but they will nonetheless continue to be experienced as separate flocks because each bird has a direction common to its flock.

Some traditional common fate examples are shown in Fig. 8.20.

We will show here movements of two different groups in the virtual, networked, space, as in Fig. 8.21, with Group 1 moving with randomised choice of horizontal (left to right) or vertical (top to bottom) individual single hops each time, and Group 2 with all nodes always making single steps from right to left. We are also assuming that group elements with different colours, as in the figure, are initially associated with network nodes via their environmental variables CONTENT, as for the previous cases.

Details of Group 1 (shown in red) movement can be understood from the following scenario:

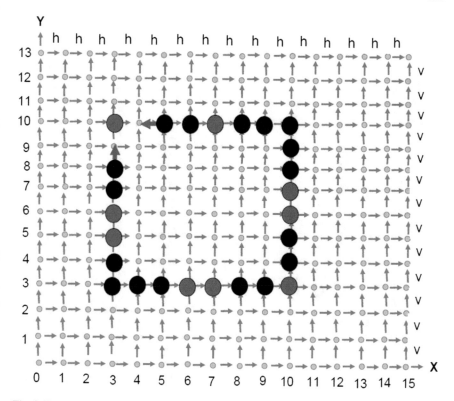

Fig. 8.18 Top left corner still to be filled

```
frontal(Color = red, Delay = timedelay);
hop_nodes(CONTENT == Color);
repeat(
  or_sequence(
      (hop_link_random(+h, -v); empty(CONTENT);
      CONTENT = Color; (BACK; CONTENT) = nil;
      sleep(Delay)),
      stay))
```

Group 2 (blue) movement is organized as follows:

```
frontal(Color = blue, Delay = timedelay);
hop_nodes(CONTENT == Color);
repeat(
    or_sequence(
    (hop_link(-h); empty(CONTENT);
    CONTENT = Color; (BACK; CONTENT) = nil;
    sleep(Delay)),
    stay))
```

Some middle situation is shown in Fig. 8.22, where the groups are intersecting with "clashes" as happen to fight for the same places in space (on nodes of the regular net in our case).

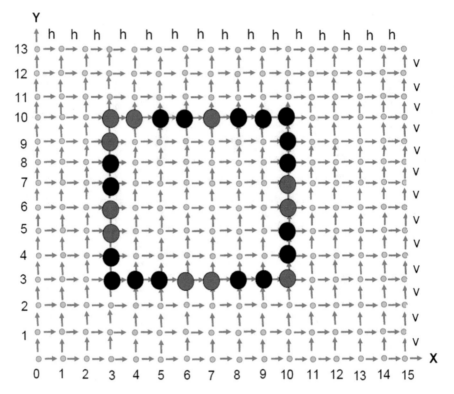

Fig. 8.19 Final solution, all gaps filled in

Fig. 8.20 Common fate traditional examples

After having gone through the intersection and clashes, the continuing collective movement via chosen directions of the groups is shown in Fig. 8.23.

We can also organize regular checking of how the groups of particular colours are moving in time, by repeatedly determining and printing coordinates of their averaged centers and the speed of shifting of these centers through the grid, as follows:

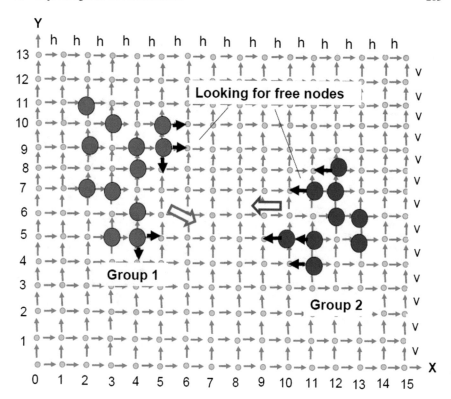

Fig. 8.21 Common fate movements: the start

```
frontal(Color) = groupcolor;
nodal(Centernew, Center = 0, Delay = timedelay,
        Speed);
sling(
   Centernew = integer_average(
                   hop_nodes(CONTENT == Color); NAME);
   Speed = distance(Centernew - Center)/Delay;
   output(Color, Centernew, Speed);
   Center = Centernew; sleep(Delay))
```

SGL allows us, similar to the considered movement in virtual, networked space, to organize any movements in real physical world too, as shown in previous publications [8–10].

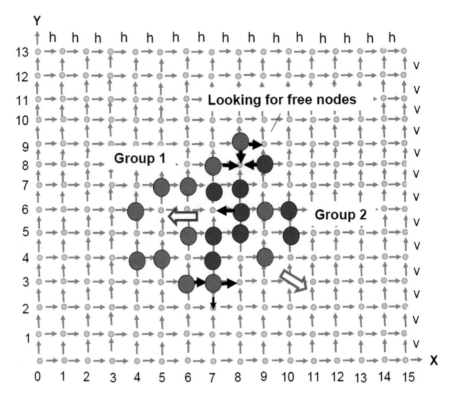

Fig. 8.22 Common fate: groups intersection

8.8 Expressing Law of Symmetry

The **Law of Symmetry** states that the mind perceives objects as being symmetrical and forming around a center point, as in traditional examples shown in Fig. 8.24. It is perceptually pleasing to divide objects into an even number of symmetrical parts, and when two symmetrical elements are unconnected the mind perceptually connects them to form a coherent shape.

Symmetry states that the viewer should not be given the impression that something is out of balance, or missing, or wrong. If an object is asymmetrical, the viewer will waste time trying to find the problem instead of concentrating on the instruction. People are accustomed to receiving information in a systematic and organized manner and will be frustrated by material that requires too much work to comprehend.

In Fig. 8.25, we consider symmetry in relation to the vertical axis for the two objects marked in different colours, and the following scenario is checking symmetry of the object represented by a given colour.

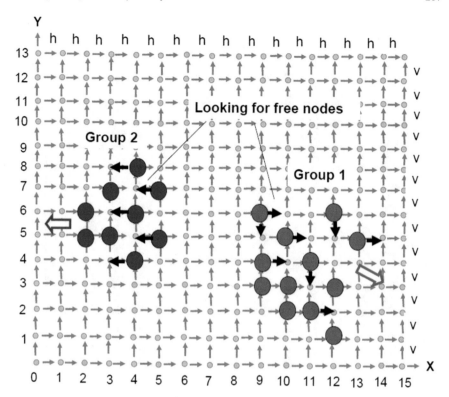

Fig. 8.23 Common fate: group movements continued

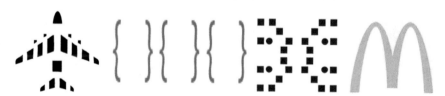

Fig. 8.24 Symmetry examples

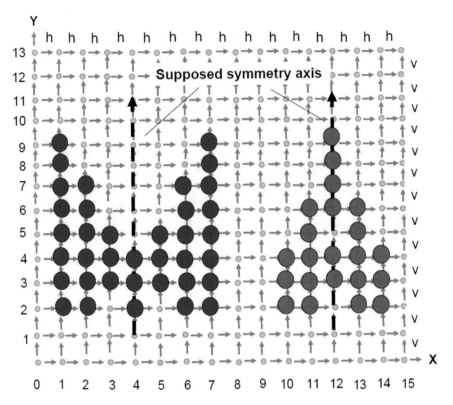

Fig. 8.25 Examples of two symmetrical objects

```
nodal(Center, Start, Step, Right, Left);
frontal(Color = color);
Center =
  integer_average(hop_nodes(CONTENT == Color); NAME);
Start = Center[1];
repeat(
  increment(Step);
  Right = sortup(hop_nodes(NAME[1] == Start + Step);
                 CONTENT == Color; NAME[2]);
  Left  = sortup(hop_nodes(NAME[1] == Start - Step);
                 CONTENT == Color; NAME[2]);
  equal(Right, Left); nonempty(Right));
empty(Right); empty(Left);
output(Color, symmetrical)
```

The printed answers for the two objects of Fig. 8.25 will be as follows (issued when initially staying in any network node or in a position outside the network):

```
blue, symmetrical
red, symmetrical
```

8.9 Figure/Ground Expression

The **Figure/Ground** concept refers to the relationship between positive elements and negative space. The idea is that the eye will separate whole figures from their background in order to understand what's being seen. It's one of the first things people will do when looking at any composition. This principle shows our perceptual tendency to separate whole figures from their background based on one or more of a number of possible variables, such as contrast, colour, size, etc. Perception of figures and grounds can change from one to the other and then back, as in Fig. 8.26.

For figures, usually *convex* rather then *concave* patterns tend to be perceived, as figures having the property that for each pair of points in the set, the line joining the points is wholly contained in the set.

A possible simple combination of figure and ground expressed by nodes of different colours is shown in Fig. 8.27.

The following scenario, for any colour detected, is checking whether corresponding whole region is convex or concave—thus classifying it as figure or ground.

```
nodal(Colors, Center); frontal(Color);
stay(hop_nodes(all);
```

Fig. 8.26 Examples of figure-ground perception

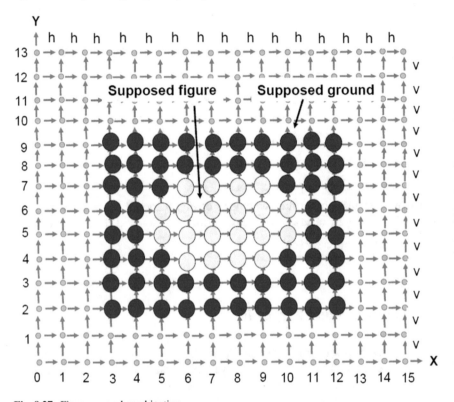

Fig. 8.27 Figure-ground combination

```
      notbelong(CONTENT, (BACK; Colors));
      (BACK; Colors) && = CONTENT);
split(Colors); Color = VALUE;
Center = integer_average(
         hop_nodes_all(CONTENT == Color); NAME);
hop_node(Center);
output(Color, if(CONTENT == Color, figure, ground))
```

Two types of output will take place for Fig. 8.27:

```
blue, ground
yellow, figure
```

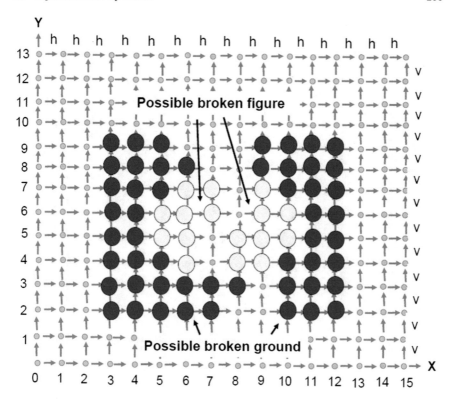

Fig. 8.28 Partitioned figure and ground areas

More detailed scenario, with figure and ground to be only fully linked areas, may be as follows, so they should not be split into pieces, like in Fig. 8.28.

```
nodal(Colors, Center); frontal(Color);
stay(hop_nodes(all);
        notbelog(CONTENT, (BACK; Colors));
        (BACK; Colors) && = CONTENT);
split(Colors); Color = VALUE; hopfirst(current);
Center = integer_average(NAME &&
    repeat(hopfirst_links(all); CONTENT == Color;
            free(NAME)));
hop_node(Center);
output(Color, if(CONTENT == Color, figure, ground))
```

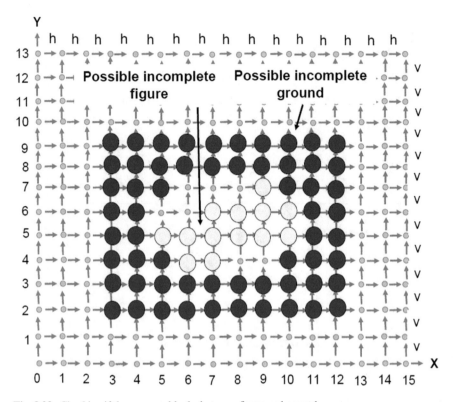

Fig. 8.29 Checking if there are no blanks between figure and ground

More details can be taken into account too, for example, with checking if figure classified by `Color1` is fully inside the region with `Color2` classified as ground, and there is no emptiness in between, so this should not happen to be as in Fig. 8.29.

```
frontal(Color1 = yellow, Color2 = blue);
if((hop_nodes_all(CONTENT == Color1);
    hop_links(all);
    notbelong(CONTENT, (Color1, Color2)),
    output(Color1, Color2, 'not real figure-ground'),
    output(Color1, Color2, 'are figure-ground'))
```

The output for the situation shown in Fig. 8.29 will be as follows:

```
Color1, Color2, not real figure-ground
```

But for the situation in Fig. 8.27, this will be issued as:

```
Color1, Color2, are figure-ground
```

We could also easily extend the techniques used in the previous scenarios for hierarchical, nested figure-ground cases similar to what is depicted in Fig. 8.30.

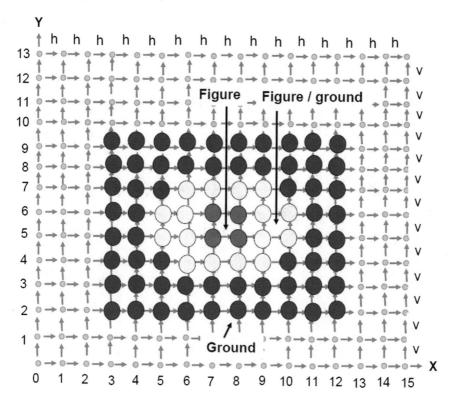

Fig. 8.30 Nested figure-ground composition

Fig. 8.31 Past experience examples

8.10 Expressing Law of Past Experience

The **Law of Past Experience** implies that under some circumstances visual stimuli are categorized according to past experience. If certain objects tend to be observed within close proximity, or small temporal intervals, the objects are more likely to be perceived together, as examples in Fig. 8.31.

Past experience is perhaps the weakest gestalt principle. In conjunction with any other principles, the other ones will dominate over the past experience principle.

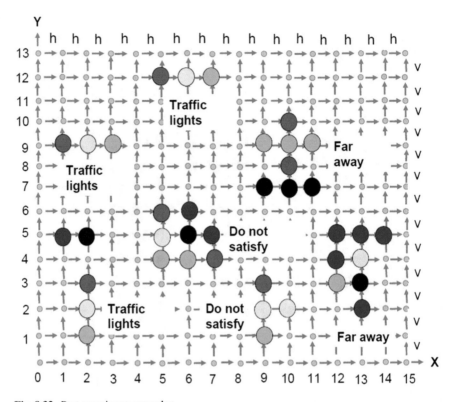

Fig. 8.32 Past experience examples

Past experience is usually unique to the individual, so it is difficult to make assumptions about how it will be perceived by others. However, there are common experiences we all share.

Having seen traffic lights throughout our lives, we expect red to mean stop and green to mean go. You see such image in Fig. 8.31 as the three common colours in proper order. That's past experience at work. We will be using this example for finding combination of these three colours in the distribution of objects shown in picture 8.32, when placed in proper order and separated from other objects.

The following scenario is finding (horizontal or vertical) sequences of three neighbouring objects with the given colours and in proper order, with these objects standing alone, i.e. not linked to other objects.

Fig. 8.33 Good gestalt examples

```
frontal(Lights = (red, yellow, green), Nodes, Link);
hop_nodes(CONTENT == withdraw(Lights));
branch(Link = +h, Link = -v);
Nodes = repeat(free(NAME); hop_link(Link);
                CONTENT == withdraw(Lights));
count(Nodes) == 3;
not(hop_nodes(Nodes); hop(links(all);
    notbelong(NAME, Nodes)); nonempty(CONTENT));
 output('Traffic lights: ' & Nodes))
```
There will be three printed results for Fig. 8.32, as follows:

```
Traffic lights: (1_9, 2_9, 3_9)
Traffic lights: (5_12, 6_12, 7_12)
Traffic lights: (2_3, 2_2, 2_1)
```

8.11 Good Gestalt Expression

The **Law of Good Gestalt** explains that elements of objects tend to be perceptually grouped together if they form a pattern that is regular, simple, and orderly. This law implies that when individuals perceive the world, they eliminate complexity and unfamiliarity so can observe a reality in its most simplistic form, with traditional examples shown in Fig. 8.33.

In Fig. 8.34 different groupings of elements are depicted. We will try to assess and classify them under the good gestalt idea, evaluating, for example, how nice they jointly use the rectangular space occupied. Other useful criteria for evaluating good gestalt qualities can be proposed and easily implemented in SGL too.

The following scenario is trying to collect groups as linked structures by asynchronous navigation of them, registering them from nodes having highest coordinate values in the group (as X_Y, represented as their names). This will block duplicates if same group registrations started in parallel from their different nodes.

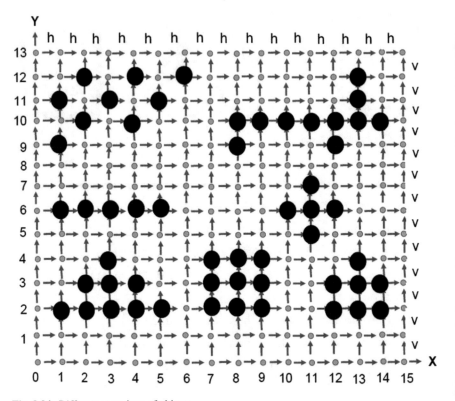

Fig. 8.34 Different groupings of objects

```
hop_nodes_all(nonempty(CONTENT));
nodal(Threshold = …); IDENTITY = NAME;
nodal(Group, Estimate, Quality, Xmin, Xmax,
      Ymin, Ymax);
contain(
  Group = NAME && repeat(
     hopfirst_links(all); nonempty(CONTENT);
     if(NAME > IDENTITY, abort); free(NAME)));
  Xmin = min(hop_nodes(Group); NAME[1]);
  Xmax = max(hop_nodes(Group); NAME[1]);
  Ymin = min(hop_nodes(Group); NAME[2]);
  Ymax = max(hop_nodes(Group); NAME[2]);
  Estimate = (Xmax - Xmin + 1) *
                (Ymax - Ymin + 1)/count(Group);
  Quality == or_sequence(
        (Estimate == 1; best),
        (Estimate < Threshold; good),
        bad);
  output(Quality & ':' & Estimate & ': ' & Group))
```

The estimates classifying gestalt quality for the found groups of Fig. 8.35 will be calculated as follows:

```
for Bad:     (14-8 + 1)*(12-9 + 1)/11 = 2.545
for Good 1:  (12-10 + 1)*(7-5 + 1)/5 = 1.8
for Good 2:  (5-1 + 1)*(5-3 + 1)/9 = 1.67
for Good 3:  (14-12 + 1)*(4-2 + 1)/7 = 1.29
for Best 1:  (5-1 + 1)*(6-6 + 1)/5 = 1
for Best 2:  (9-7 + 1)*(4-2 + 1)/9 = 1
```

8.12 Expressing Connectedness or the Law of Unity

The **Uniform Connectedness (Law of Unity)**. Elements that are connected by uniform visual properties, such as colour or direct links, are perceived to be more related than elements that are not connected. The principle of uniform connectedness is the most recent addition to the principles referred to as Gestalt principles of perception. It asserts that elements connected to one another by uniform visual properties are

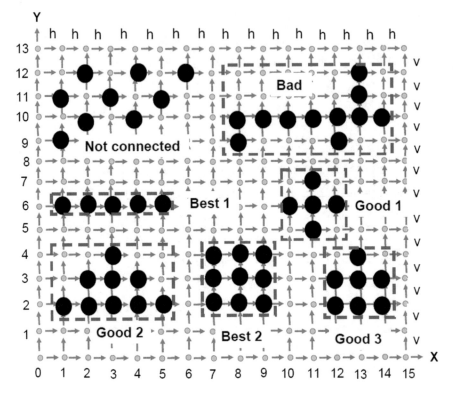

Fig. 8.35 Good gestalt estimates

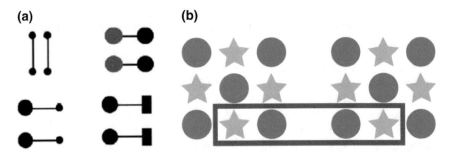

Fig. 8.36 Connectedness examples: **a** connecting lines; **b** common region

perceived as a single group or chunk and are interpreted as being more related than elements that are not connected. There are two basic strategies for applying uniform connectedness in a design: *common regions and connecting lines* (related examples in Fig. 8.36).

Common regions are formed when edges come together and bound a visual area, grouping the elements within the region. *Connecting lines* are formed when an

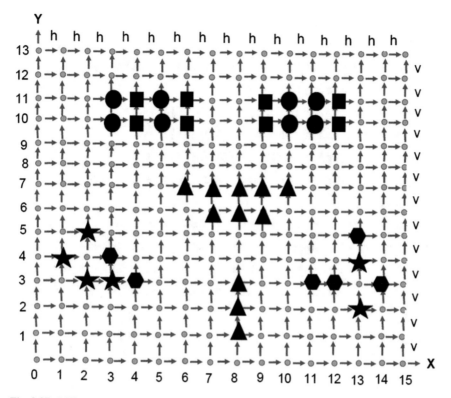

Fig. 8.37 Different groups of elements to be connected

explicit line joins elements, grouping the connected elements. This technique is often used to connect elements that are not otherwise obviously grouped) or to imply a sequence.

Having placed different objects on the regular network grid as in Fig. 8.37, we will be showing here *common region* or *connecting lines* techniques to join different elements within separate groups.

We will consider the following cases and their expression in SGL for the set of objects shown in the figure.

- *Case 1*: Outlining minimal common region containing all circle and square objects using a red line surrounding it, with the region having square shape.

```
nodal(Group, Xmax, Xmin, Ymax, Ymin);
frontal(Start, Fin, Link)
Group = (hop_nodes(belong(CONTENT, (circle,
          square))); NAME);
Xmax = max(split(Group); VALUE[1]) + 1;
Xmin = min(split(Group); VALUE[1]) - 1;
Ymax = max(split(Group); VALUE[2]) + 1;
Ymin = min(split(Group); VALUE[2]) - 1;
parallel(
  (Start = Xmin_Ymin; Fin = Xmax_Ymin; Link = +h),
  (Start = Xmin_Ymax; Fin = Xmax_Ymax; Link = +h),
  (Start = Xmin_Ymin; Fin = Xmin_Ymax; Link = +v),
  (Start = Xmax_Ymin; Fin = Xmax_Ymax; Link = +v));
hop_node(Start);
repeat(hop_link(Link); nonequal(Fin, NAME);
          stay_linkup(link(red), BACK))
```

- *Case 2 and Case 3*: Linking all directly neighboring triangles into groups.

```
hop_nodes(CONTENT == triangle);
hop(links(all), node(CONTENT == triangle));
NAME < PREDECESSOR; linkup(link(red), BACK)
```

- *Case 4 and Case 5*: Linking all directly neighboring (i.e. at length 1) or neighboring to neighboring (i.e. at length 2) nodes with stars or hexagons into groups by spanning trees starting from nodes with maximum value as their name.

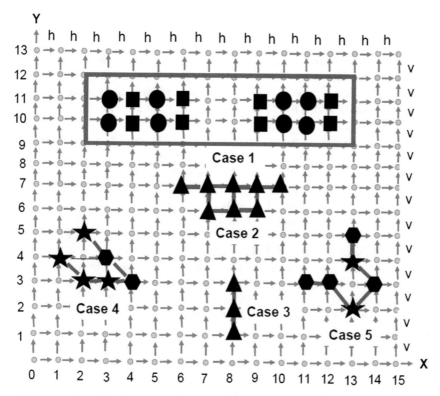

Fig. 8.38 Results of establishing common region or direct internode connections

```
frontal(objects) = (star, hexagon);
nodal(Last);
hopfirst_nodes(belong(CONTENT, Objects));
IDENTITY = NAME;
contain(
  repeat(
    Last = NAME;
    or_sequence(hopfirst_links(any),
                2_hopfirst_links(any);
                belong(CONTENT, Objects));
    if(NAME > IDENTITY, abort);
    stay_linkup(link(red), node(Last))))
```
The results for all considered cases are depicted in Fig. 8.38.

8.13   Conclusion

Although Gestalt theory was born as a very general approach showing the unique capability of perceiving the whole first an then treating parts as second, in the context of the whole, most of works on gestalt were conducted in the area of perception of images, mainly visual, by localized human brain. In our true belief, the theory and practice of gestalt can be effectively used in much broader sense and scale, and especially for holistic vision, comprehension and proper management of large distributed dynamic systems of most different natures,—from economy to ecology to security to defense and, of course, to social life and social systems in general.

In this chapter we conducted experiments to show how basic gestalt laws can work in fully distributed manner, by first, placing some related images on a canvas of regular networks which can be arbitrarily distributed in both virtual and physical spaces and can span the whole world. And by second, by presenting high-level scenarios in SGL that can spread over, cover, and match these images with showing their proper perception an understanding which can be done without any central resources. These scenarios appeared to be very compact and transparent, confirming potential applicability of gestalt laws for large distributed systems. The latter can definitely include large social systems and their networks, and this area is planned for a further investigation and research.

The conducted experiments on expressing gestalt laws in distributed systems had another goal too: to test the very principles underlying SGT and SGL for their use in untraditional and complex area, which gave positive answers.

Other works related to SGT and SGL and their previous versions, including numerous applications in different areas [12–21] were always inspired and guided by holistic ideas stemming from gestalt psychology and theory, also strengthened by the author's practical perception of incomplete spatial images as amateur painter and sculptor. The influenced holistic thinking and world perception often contradicted traditional logic [12], also seeing the world (our mind including, as in [22]) as consisting of numerous "agents" which in their interaction can hopefully produce something whole and great—and this does not work properly in many cases. (The author often recollects his discussions years ago and in the US on that matter with the author of [22] Marvin Minsky.)

References

1. M. Wertheimer, Untersuchungen zur Lehre von der Gestalt, I: Prinzipielle Bemerkungen. Psychologische Forschung (1922)
2. W. Köhler, Die physischen Gestalten in Ruhe und im stationären Zustand. Eine naturphilosophische Untersuchung. Braunschweig. Germany: Friedr. Vieweg und Sohn (1920)
3. K. Koffka, *Principles of Gestalt Psychology* (Lund Humphries, London, U.K., 1935)
4. J. Wagemans, J.H. Elder, M. Kubovy, S.E. Palmer, M.A. Peterson, M. Singh, R. von der Heydt, A century of gestalt psychology in visual perception: I. Percept. Grouping Figure-Ground Organ Psychol Bull **138**(6), 1172 (2012)

5. S. Bradley, Design principles: visual perception and the principles of gestalt. Smashing Mag. (2014). https://www.smashingmagazine.com/2014/03/design-principles-visual-perception-and-the-principles-of-gestalt/#summary-of-gestalt

6. P. Sapaty, in *Gestalt-Based Ideology and Technology for Spatial Control of Distributed Dynamic Systems*, International Gestalt Theory Congress, 16th Scientific Convention of the GTA, University of Osnabrück, Germany, 26–29 March 2009

7. P. Sapaty, in *Gestalt-Based Integrity of Distributed Networked Systems* (SPIE Europe Security + Defence, bcc Berliner Congress Centre, Berlin Germany, 2009)

8. P. Sapaty, *Mobile Processing in Distributed and Open Environments* (Wiley, New York, 1999)

9. P. Sapaty, *Ruling Distributed Dynamic Worlds* (Wiley, New York, 2005)

10. P. Sapaty, in *Managing Distributed Dynamic Systems with Spatial Grasp Technology* (Springer, 2017)

11. M.B. Salas, in *A Unified Whole: Gestalt and the Group* (2016). https://www.linkedin.com/pulse/unified-whole-gestalt-group-may-bethzaida-salas/

12. P.S. Sapaty, in *Logic Flow in Active Data*, VLSI for Artificial Intelligence and Neural Networks (1991), pp. 79–91

13. P. Sapaty, in *Providing Over-Operability of Advanced ISR Systems by a High-Level Networking Technology*, SMI's Airborne ISR (Holiday Inn Kensington Forum, London, United Kingdom, 2015)

14. P.S. Sapaty, Over-operability in distributed simulation and control. MSIAC's M&S J. Online, Winter Issue, **4**(2). Alexandria, VA, USA (2002)

15. P.S. Sapaty, in *The World as Distributed Brain with Spatial Grasp Paradigm*, Intelligent Systems for Science and Information, Vol. 542 of the Series Studies in Computational Intelligence (2014), pp. 65–85

16. P.S. Sapaty, in *Global Electronic Dominance*, 12th International Fighter Symposium (Grand Connaught Rooms, London, UK, 2012)

17. P. Sapaty, in *Providing Global Awareness in Distributed Dynamic Environments*, International Summit ISR (London, 2013)

18. P.S. Sapaty, Towards global goal orientation, robustness and integrity of distributed dynamic systems, J. Int. Relat. Diplomacy **4**(6) (2016)

19. P.S. Sapaty, in *A Distributed Processing System*. European Patent No. 0389655. European Patent Office (1993)

20. P.S. Sapaty, in *Distributed Technology for Global Control*, Informatics in Control, Automation and Robotics, Vol. 37 of the Series Lecture Notes in Electrical Engineering (2009), pp. 3–24

21. P.S. Sapaty, in Grasping Spatial Integrity in Distributed Unmanned Systems, book chapter in Informatics in Control Automation and Robotics, Vol. 85 of the series Lecture Notes in Electrical Engineering, 2011, pp. 79–97

22. M. Minsky, In *The Society of Mind, Simon & Schuster* (1986)

Chapter 9
Conclusions

9.1 Summary of the Main Results

The book briefed the developed Spatial Grasp Technology for high-level parallel and distributed creation, analysis, and management of large distributed systems of different natures, which may be virtual (electronic), physical, or combined. Such systems may have no borders and can cover any territories, the whole world including. The solutions in such systems are organized in a holistic and integral way, by self-spreading and self-matching recursive scenarios-patterns spatially grasping them. This, in some sense, resembling our mental activity and flexibility with scanning and perception of distributed spaces and images in them, which is quite different from the existing models and approaches based on communicating parts-agents and their interactions.

This particular book is devoted to a relatively new application area, not explored in detail in the previous publications, and namely, large distributed social systems and their expression as social networks, which may have any structure and numerous nodes and links connecting them. The latest full version of Spatial Grasp Language, as the core of the approach developed, is presented in detail, being particularly suitable for dealing with large social systems.

Numerous scenarios have been shown in SGL and explained in this book, related to basic operations on general graphs and networks, dealing with classical networking centrality issues and clustering, offering peculiar operations on specialized types of social, business, and industrial networks, as well as very large and unknown networks. Shown also was the technology's applicability in such areas as human-robotic teaming and driverless transport, which are becoming inseparable parts and features of advanced societies. We also succeeded in demonstrating in SGL of how known gestalt theory laws could work with distributed spatial images too, with the resultant importance in many areas, social systems including. The design and description in SGL of numerous solutions for social systems provided us with new experience of

P. S. Sapaty, *Holistic Analysis and Management of Distributed Social Systems*, Studies in Systems, Decision and Control 184, https://doi.org/10.1007/978-3-030-01830-6_9

dealing with complex networking problems, which we plan to use in subsequent research and publications.

The remaining concluding sections will be (a) summarizing the essence of the approach developed on a very simple but representative networking example also explaining why the approach developed has *spatial grasp* in its name; (b) providing guidelines for quickest, simplified SGL implementation with further extensions if needed; (c) naming planned areas for further research, design, and implementation.

9.2   A Concluding Explanatory Example

We have shown many networking scenarios in this book in SGL, but hundreds more, covering very different fields in distributed network systems, were written (in SGL, also in previous language variants, WAVE including), tested in real networks and published. Related summaries can be found in [1–3].

All these were expressed in the language with very simple syntax but pursuing quite different, even unusual, ideology and psychology of programming in it, where highly parallel and fully distributed algorithms were not based on traditional communication parts or agents but rather representing active holistic patterns spreading through and matching arbitrary network topologies.

These patterns are extremely compact, transparent and simple, due to expressing only top semantics of arbitrary networking solutions, with most of traditional programming routines (often up to 99% in programs written in languages like C or Java) are effectively hidden inside the intelligent, networked SGL interpreter; they do not need showing them each time explicitly.

So, there are *no traditional parts-agents* and their communications in the developed programming paradigm and its SGL language, as these are effectively shifted to something conceptually and organizationally much lower. Actually, SGL interpreters and their communications *are the interacting agents*, but only representing the universal implementation machine which is tasked and governed on top by gestalt-related, spatial and indivisible SGL scenarios, which *are the application programs* in our case.

These scenarios are often self-changeable and very flexible, reflecting our mental capabilities in scanning, perception, planning and changing of spatial structures, often having the great advantage of being highly parallel in nature and thus superseding human mind and brain capabilities in this respect.

In an attempt to summarize and emphasize at the end of this book some main features of the approach developed, will be showing here how to find all simple paths (i.e. without repeating nodes) from some node A to another one B in arbitrary network (will be using the one of Fig. 9.1). This solution will be repeating a similar example of Chap. 4 but with more explanations and general conclusions.

The related SGL scenario may be as follows:

```
hop(A); frontal(Path);
```

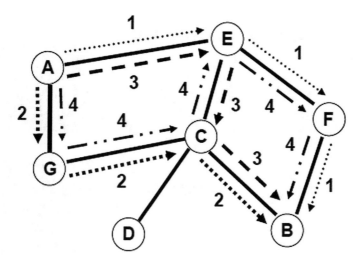

Fig. 9.1 All simple paths from node A to node B

```
repeat(append(Path, NAME);
        if(NAME == B, quit(output(Path)));
        hop(links(all)); notbelong(NAME, Path))
```

In this scenario, we are first landing directly on node A, then driving directly, independently, and in parallel through all links from current nodes to neighboring nodes not reached before. The simple passed paths are being accumulated during moving through the network in frontal, moving variables Path (individual for each thread) while providing final outputs directly in node B upon reaching it. The construct covered by rule repeat is also moving through the network while being replicated if more than one non-visited neighboring node occurs, same as frontal variables Path.

We will be having the following outputs in node B (reflected in Fig. 9.1 by different types of arrows, which are also additionally numbered):

```
(A,E,F,B),  (A,E,C,B),  (A,G,C,B),  (A,G,C,E,F,B)
```

With slight modification of the above SGL scenario, if needed, all paths found can be returned to node A and independently issued in it, as follows:

```
hop(A); frontal(Path);
output(
  repeat(append(Path, NAME);
          if(NAME == B, blind(Path));
          hop(links(all)); notbelong(NAME, Path)))
```

In this variation, the hidden, internal tracking system of distributed SGL interpretation will be automatically employed to return all collected paths from the final node B (reached repeatedly by different threads) whatever remote it might occur.

In Fig. 9.2 the tracking structure of distributed interpretation of this particular scenario is shown, which allowed us to return the accumulated paths by different threads that reached final node B to the starting node A and issue them all there.

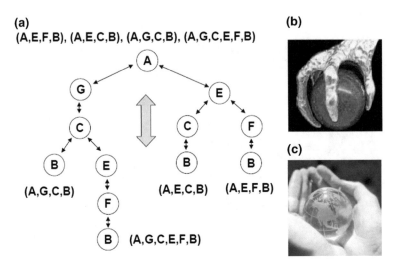

Fig. 9.2 Track-based distributed SGL implementation. **a** The tracking structure. **b–c** Symbolic analogies

Despite simplicity of the example shown, it nevertheless highlights and summarizes some main features of the approach offered and methods of solving tasks with it on general graphs and networks. This is parallel movement through them in *forward* or covering and exploration direction, and backward or *echo* parallel propagation for collecting potentially remote data and summarizing control states for making proper decisions. Any combinations of these forward and echo schemes can be possible within generally recursive distributed SGL-based algorithms.

This combination of the forward and echo operations on general networks (in physical spaces too) provided by SGL has prompted using the word **grasp** in the model, language, and technology definitions, whereas previous language and technology versions were mostly using the word **wave**. These wave versions operated without using internal tracks for returning remotely obtained data and the language organizations were non-recursive and therefore much simplified.

Very symbolic analogies of the developed Spatial Grasp Technology which can intellectually grasp any distributed systems, world-wide including, and bring back whatever may be needed, are shown in Fig. 9.2b for possible "aggressive" applications, and in Fig. 9.2c for "benign" or friendly ones.

9.3 Minimum Subset for Quick Implementation

Described in Chap. 3 the "full-bodied" SGL, as a result of decades of dealing with a variety of distributed systems, both civil and military, and in different countries, has a variety of very useful constructs which emerged from the development, optimization and simplification of numerous network navigation and processing scenarios, known from the past as "waves" with same named language WAVE [4–13]. But SGL has very simple and universal recursive syntax of its core subset, which can be expressed as follows:

Grasp ➜ *constant | variable | rule* ({*grasp ,* })
variable ➜ *frontal | nodal | environmental*
rule ➜ *movement | creation | echoing | verification | assignment | advancement | branching*

Moreover, the language *rules* can be often substituted for many applications just by delimiters, operations, and control symbols common in traditional languages, further simplifying the starting implementation which can be later extended for the growing sets of tasks in a particular area. Such basic subset can be easily implemented by a couple of system programmers even within usual university environments, as was done in the past in different countries, first in Lisp [14–16], and then in C [17–19]. In these projects the author was usually serving as team's "playing coach", distributed algorithms designer and top scenario programmer, as well as supervisor of related M.Sc. and Ph.D. projects.

Implementation of the language interpreters can be also be integrated with any existing internet, popular media, robotic, and command and control systems. Full language version presented in Chap. 3 can be readily implemented for extended applications too, in full cooperation with the author.

9.4 Future Plans

Further plans, after this book publication, are aimed at the development and use of SGT and its further modifications in different directions, and especially related to very large distributed dynamic systems—economic, industrial, business and, of course, social ones including.

9.4.1 Economic Systems Research

Of particular interest will be the Japanese economy [20]. We have analyzed GDP growth in 2018 for top three world economies, with the latest figures being extremely interesting: US 4.1% [21]; China 6.7% [22]; Japan only 1.1%, and in 2019 even expected down to 1% [23]. Why so low in Japan? As a result of investigation, the main problems for the economy in Japan look like being liked with its outdated structural organization, and especially with Keiretsu networks, which are ageing and losing flexibility.

A Keiretsu system [24–26] (see also Chap. 6) is a set of companies with interlocking business relationships and shareholdings. The major *keiretsu* were each centered on one bank which acted as both monitoring and as emergency bail-out entity. As mentioned in Chap. 6, there are two types of *keiretsu*: vertical and horizontal. *Vertical keiretsu* illustrates the organization and relationships within a company, while *horizontal keiretsu* shows relationships between entities and industries, centered on a bank and trading company. Both are complexly woven together and sustain each other.

There is a rising strong feeling in the business community that keiretsu system is loosening its effectiveness, and this, in our opinion, can become a *new field of scientific research* and investigation. With the ideology and technology developed we can also organize runtime simulation and management in the economy, offering a sort of distributed spatial brain [27] overseeing and optimizing large economic systems as a whole. The work already started in this direction, which is also planned to be part of an international project dealing with real social, economic, business, and industrial systems and networks, keiretsu ones including, with new publications on the results obtained, further books too.

We also hope that the gained former experience within world biggest Distributed Interactive Simulation (DIS) project (defense-oriented, headquartered in Orlando, Fl) under the previous versions of the technology and language [28–32] can be particularly useful for simulation and management of large economic and industrial systems.

9.4.2 Providing Integrity of Very Large Systems

The future plans also include a much deeper analysis of the classical works on general systems theory [33–38] and system dynamics [39–42] with possible reconsidering their sense, value, and applications from the holistic viewpoint, and especially after our attempts in this book to extrapolate known gestalt principles [43–46] to large distributed systems. Also to be reassessed and further developed our related previous works on holistic analysis and management of distributed dynamic systems, global electronic dominance, global awareness in dynamic environments, wholeness and integrity of distributed systems, global goal orientation, systems robustness and integrity, and others [47–54].

The SGT can effectively influence the whole distributed world by converting it into a sort of spatial brain with the help of *Social Analysis Devices* (SAD) as extended SGL interpreters [55], which can be implanted into sensitive points of social tissue (on agreements or in a stealth manner for certain, say, law and order applications) while communicating with each other. Dynamic SAD networks can be embedded into distributed social networks tissue as its integral part, and may be very large, up to millions to billions of nodes, both stationary and mobile. They can cooperatively collect and extract important, including peculiar and sensitive information on social events, feelings, and aspirations, also discover and analyze different kinds of distributed social infrastructures, which may be benign or malicious.

This is achievable, as shown in this and the previous books, by self-evolving, self-growing, self-replicating and self-spreading patterns written in a special recursive language, which can be applied from any SAD modules while creating higher-level holistic operational and awareness infrastructures dynamically covering and matching the social areas of interest in a globally controlled mode, which can function without any centralized resources.

By using distributed SAD networks directly accessing the freshest social information in numerous places, it will be possible to analyze, simulate and predict different developments of social systems under realistic or hypothetical circumstances, also conduct controlled local and global social experiments. This can be useful for solving a variety of problems emerging on national and international levels—from welfare to security to prosperity to stability to defense to international relations and diplomacy, in this highly dynamic and unpredictable century.

9.4.3 New Patenting, Implementation, and Marketing

Patenting the latest version of the technology developed (to succeed the previous, old, patent [56]) is on the agenda, as effective distributed software or hardware implementation of SGL-type languages for solving problems in spatial pattern-matching mode, without any centralized resources, may contain powerful and not yet reported/published mechanisms worth of new international patents. The technology can be readily ported on any new platforms, including embedment into modern media systems or email services, which can be always be done with the author's assistance. Its marketing in both civil and defence areas is in plans too, as the tech *allows us to provide holistic and integral solutions in a variety of distributed environments—from economy to welfare to ecology to culture to education to international relations to security to battlefields, which may be often superior to other existing models and approaches.*

References

1. P. Sapaty, *Mobile Processing in Distributed and Open Environments* (Wiley, New York, 1999)
2. P. Sapaty, *Ruling Distributed Dynamic Worlds* (Wiley, New York, 2005)
3. P. Sapaty, *Managing Distributed Dynamic Systems with Spatial Grasp Technology* (Springer, Berlin, 2017)
4. P.S. Sapaty, A wave language for parallel processing of semantic networks. Comput. Artif. Intell. **5**(4), 289–314 (1986)
5. P.S Sapaty, The WAVE-0 language as a framework of navigational structures for knowledge bases using semantic networks, in *Proceedings of USSR Academy of Sciences*. Technical Cybernetics, No. 5 (1986) (in Russian)
6. P.S. Sapaty, I. Kocis, A parallel network wave machine, in *Proceedings of 3rd International Workshop PARCELLA'86* (Akademie-Verlag, Berlin, 1986)

7. P.S. Sapaty, The wave approach to distributed processing of graphs and networks, in *Proceedings of International Working Conference Knowledge and Vision Processing Systems*, Smolenice, November 1986
8. P.S. Sapaty, A wave language for parallel processing of semantic networks. Comput. Artif. Intell. **5**(4), 289–314 (1986)
9. P.S. Sapaty, The WAVE-1: a new ideology and language of distributed processing on graphs and networks. Comput. Artif. Intell. **5** (1987)
10. P.S. Sapaty, WAVE-1: a new ideology of parallel processing on graphs and networks, in *Proceedings of International Conference on Frontiers in Computing* (Amsterdam, 1987)
11. Sapaty, P.S. WAVE-1: A new ideology of parallel processing on graphs and networks, Future Gener. Comput. Syst. **4** (North-Holland) (1988)
12. P.S. Sapaty, The WAVE machine project, in *Proceedings of IFIP Workshop on Silicon Architectures for Neural Nets, St. Paul de Vence*, France, 28–30 November 1990
13. P.S. Sapaty, *A Brief Introduction to the WAVE Language*, Report No. 3/93, Faculty of Informatics, University of Karlsruhe, 1993
14. P. Sapaty, S. Varbanov, A. Iljenko, The WAVE model and architecture for knowledge processing, in *Proceedings of Fourth International Conference on Artificial Intelligence and Information-Control Systems of Robots*, Smolenice (1987)
15. P.S. Sapaty, S. Varbanov, M. Dimitrova, Information systems based on the wave navigation techniques and their implementation on parallel computers, in *Proceedings of International Working Conference on Knowledge and Vision Processing Systems*, Smolenice, November 1986
16. S. Varbanov, P.S. Sapaty, An information system based on the wave navigation techniques, in *Abstracts of the International Conference on AIMSA'86*, Varna, Bulgaria, 1986
17. P.S. Sapaty, P.M. Borst, *Distributed Computing with WAVE*, Technical Report, Dept. Electronic & Electrical Eng, University of Surrey, May 1995
18. P.S. Sapaty, P.M. Borst, *An Overview of the WAVE Language and System for Distributed Processing in Open Networks*, Technical Report, Dept. Electronic & Electrical Eng, University of Surrey, June 1994
19. P.M. Borst, An Architecture for Distributed Interpretation of Mobile Programs, PhD Dissertation, University of Karlsruhe, 2001, also published as a book by Herbert Utz Verlag Gmbh (2002)
20. Economy of Japan. https://en.wikipedia.org/wiki/Economy_of_Japan
21. United States GDP Growth Rate, Trading Economics, Jul 2018. https://tradingeconomics.com/united-states/gdp-growth
22. China GDP Annual Growth Rate, Trading Economics, Jul 2018. https://tradingeconomics.com/china/gdp-growth-annual
23. Japan Economic Look, Focus Economics, Jul 245, 2018. https://www.focus-economics.com/countries/japan
24. J. Grabowiecki, *Keiretsu Groups: Their Role in the Japanese Economy and a Reference Point (or a paradigm) for Other Countries*, V.R.F. Series, No. 413, Mar 2006. http://www.ide.go.jp/library/English/Publish/Download/Vrf/pdf/413.pdf
25. J.R. Lincoln, M. Shimotani, Whither the Keiretsu, Japan's Business Networks? How Were They Structured? What Did They Do? Why Are They Gone? IRLE Working Paper No. 188–09 (2009). http://irle.berkeley.edu/workingpapers/188-09.pdf
26. A. Takeishi, Y. Noro, Keiretsu Divergence in the Japanese Automotive Industry: Why Have Some, But Not All, Gone? CEAFJP Discussion Paper Series 17–04, CEAFJPDP, August 2017. http://ffj.ehess.fr/upload/Discussion/CEAFJPDP-17-04.pdf
27. P.S. Sapaty, The World as Distributed Brain with Spatial Grasp Paradigm, Book Chapter in Intelligent Systems for Science and Information, Vol. 542 of the series Studies in Computational Intelligence (2014), pp. 65–85
28. P. Sapaty, M.J. Corbin, S. Seidensticker, Mobile intelligence in distributed simulations, in *Proceedings of 14th Workshop on Standards for the Interoperability of Distributed Simulations*, IST UCF, Orlando, FL, March 1995

29. P.S. Sapaty, P.M. Borst, M.J. Corbin, J. Darling, Towards the intelligent infrastructures for distributed federations, in *Proceedings of 13th Workshop on Standards for the Interoperability of Distributed Simulations*, IST UCF, Orlando, FL, Sept 1995, pp. 351–366
30. P.S. Sapaty, M.J. Corbin, P.M. Borst, Mobile WAVE programming as a basis for distributed simulation and control of dynamic open systems, Report at the 4th UK SIWG National Meeting, SGI Reality Centre, Theale, Reading, 11 October 1994
31. P.S. Sapaty, M.J. Corbin, P.M. Borst, Towards the development of large-scale distributed simulations, in *Proceedings of 12th Workshop on Standards for the Interoperability of Distributed Simulations*, IST UCF, Orlando, FL, March 1995, pp. 199–212
32. P.S. Sapaty, A new technology for integration, simulation, and testing of distributed dynamic systems, in *NATO Proceedings Integration of Simulation with System Testing*, RTO-MP-083, AC/323(SCI-083)TP/43, June 2002
33. L. von Bertalanffy, *General System Theory Foundations, Development, Applications*. George Braziller, New York, 1968
34. L. von Bertalanffy, The theory of open systems in physics and biology. Science **111**(2872), 23–29 (1950)
35. L. von Bertalanffy, Theoretical models in biology and psychology, J. Pers. **20**(1) (1951)
36. D. Pouvreau, M. Drack, On the history of Ludwig von Bertalanffy's "General Systemology", and on its relationship to cybernetics. Part I: elements on the origins and genesis of Ludwig von Bertalanffy's "General Systemology". Int. J. Gen Syst **36**(3), 281–337 (2007)
37. M. Drack, Ludwig von Bertalanffy's organismic view on the theory of evolution, J. Exp. Zool. (Mol. Dev. Evol.) **324B**, 77–90 (2015)
38. D. Pouvreau, On the history of Ludwig von Bertalanffy's "general systemology", and on its relationship to cybernetics—Part II: contexts and developments of the systemological hermeneutics instigated by von Bertalanffy. J. Gen. Syst. **43**(2), 172–245 (2014)
39. J. Forrester, *Urban Dynamics*. Pegasus Communications (1969)
40. J. Forrester, *World Dynamics* (Wright-Allen Press, Cambridge, MA, 1971)
41. J. Forrester, *Industrial Dynamics* (Pegasus Communications, Waltham, MA, 1961)
42. J. Forrester, Learning Through System Dynamics as Preparation for the 21st century (2009)
43. J. Wagemans, J.H. Elder, M. Kubovy, S.E. Palmer, M.A. Peterson, M. Singh, R. von der Heydt, A century of gestalt psychology in visual perception: I. Perceptual grouping and figure-ground organization. Psychol. Bull. **138**(6), 1172 (2012)
44. M. Wertheimer, Untersuchungen zur Lehre von der Gestalt, I: Prinzipielle Bemerkungen. Psychologische Forsch (1922)
45. W. Köhler, Die physischen Gestalten in Ruhe und im stationären Zustand. Eine naturphilosophische Untersuchung (Friedr. Vieweg und Sohn, Braunschweig, 1920)
46. K. Koffka, *Principles of Gestalt psychology* (Lund Humphries, London, U.K., 1935)
47. P.S. Sapaty, Global electronic dominance, in *12th International Fighter Symposium*, 6th–8th November 2012, Grand Connaught Rooms, London, UK
48. P. Sapaty, Providing global awareness in distributed dynamic environments, in *International Summit ISR*, London, April 16–18, 2013
49. P. Sapaty, Ruling distributed dynamic worlds with spatial grasp technology, in *Tutorial at the INTERNATIONAL SCIENCE and Information Conference 2013 (SAI)*, 7–9 October 2013, London, UK
50. P.S. Sapaty, Towards wholeness and integrity of distributed dynamic systems, J. Comput. Sci. Syst. Biol. **9**, 3 (2016)
51. P.S. Sapaty, Towards global goal orientation, robustness and integrity of distributed dynamic systems. J. Int. Relat. Diplomacy **4**(6) (2016)
52. P. Sapaty, Gestalt-based integrity of distributed networked systems, SPIE Europe Security+Defence, bcc Berliner Congress Centre, Berlin Germany (2009)
53. P. Sapaty, Distributed technology for global dominance, in *Proceedings of International Conference Defense Transformation and Net-Centric Systems 2008*, as part of the SPIE Defense and Security Symposium, 16–20 March 2008, World Center Marriott Resort and Convention Center, Orlando, FL, USA (Proceedings of SPIE – Volume 6981, Defense Transformation and Net-Centric Systems 2008, Raja Suresh, Editor, 69810T, Apr. 3, 2008)

54. P. Sapaty, Distributed technology for global dominance, keynote lecture, in *Proceedings of the Fifth International Conference in Control, Automation and Robotics ICINCO 2008*, 11–15 May 2008, Funchal, Madeira, Portugal
55. P. Sapaty, Holistic analysis and management of distributed social systems, in *5th International Conference on Big Data Analysis and Data Mining*, June 20–21, Rome, Italy, also in Journal of Computer Engineering & Information Technology, vol 7 (2018). https://www.scitechnol.co m/conference-abstracts-files/2324-9307-C1-020-002.pdf
56. P. Sapaty, A Distributed Processing System, European Patent No. 0389655, Publ. 10.11.93, European Patent Office, Munich (1993)

Printed in the United States
By Bookmasters